Cleaning Pakistan's Air

DIRECTIONS IN DEVELOPMENT
Energy and Mining

Cleaning Pakistan's Air

Policy Options to Address the Cost of Outdoor Air Pollution

Ernesto Sánchez-Triana, Santiago Enriquez, Javaid Afzal, Akiko Nakagawa, and Asif Shuja Khan

THE WORLD BANK
Washington, D.C.

In memory of Gajanand Pathmanathan

Contents

Figures

Map

Tables

emission standards, and technology- and performance-based standards. However, Pakistan's policy makers face major obstacles, including limited financial, human, and technical resources, and can pursue only a few AQM interventions at the same time. Hence, priorities need to be determined, and this book includes the following recommendations:

In the short term, Pakistan's AQM should give highest priority to reducing pollutants linked to high morbidity and mortality: fine particulate matter (and precursors like sulfur oxides and nitrogen oxides) from *mobile* sources. A second-level short-term priority could be on fine particulate matter, sulfur oxides, and emissions of lead and other toxic metals from *stationary* sources. An important medium-term priority should be developing mass transportation in major cities, controlling traffic, and restricting private cars during high-pollution episodes. A long-term priority could be taxing hydrocarbons, based on their contribution to greenhouse gases.

Pakistan faces daunting problems stemming from urban air pollution's adverse effects on Pakistanis' health, quality of life, economy, and environment. The underlying goal of this book is to facilitate and stimulate further sharing of information regarding those harms and to provide a research-based framework for bringing about improved air quality management in Pakistan.

John Henry Stein
Sector Director
Sustainable Development Department
South Asia Region
World Bank

Acknowledgments

This book is a product of a core team that included Ernesto Sánchez-Triana (TTL), Javaid Afzal, Santiago Enriquez, Asif Shuja Khan, Bjorn Larsen, John Skelvick, Manuel Dall'Osto, Akiko Nakagawa, Cecilia Belita, Rahul Kanakia, Mosuf Ali, and Hammad Raza Khan. The extended team included Asif Faiz, Anil Markandya, Sameer Akbar, Luis Sánchez Torrente, Mohammad Omar Khalid, and Badar Munir Ghauri. Sada Hussain and Afzal Mahmood provided administrative support.

The peer reviewers for the study were Helena Naber, Daniel Mira-Salama, John Allen Rogers, and Yewande Awe (World Bank); Jack Ruitenbeek (IUCN Switzerland); and Leonard Ortolano (Stanford University). The team also benefited from the comments of Luis Alberto Andres, Dan Biller, Catherine Revels, Karin Kemper, and Gajanand Pathmanathan. Several colleagues also provided helpful advice and detailed contributions, particularly Asif Faiz, Eugenia Marinova, Anna O'Donnell, Mario Picon, and Maria Correia. The authors are also thankful for the support of the World Bank management team and the Pakistan country office, including Rachid Benmessaoud (Country Director), John Henry Stein, Karin Kemper, Gajanand Pathmanathan, Herbert Acquay, and Bernice Van Bronkhorst. The authors extend their sincere thanks and appreciation to Cecilia Belita, Marie Florence Elvie, Afzal Mahmood, and Sada Hussain for their administrative support. Editorial and manuscript preparation was provided by Stan Wanat.

The Government of Pakistan, mainly through the Pakistan Environmental Protection Agency and the Ministry of Climate Change, provided key feedback during the preparation of the study and participated actively in the production of diverse parts of the book. The team received significant contributions and feedback from participants in a workshop held during May 2010 in Islamabad. Participants included Mr. Kamran Ali Qureshi, Additional Secretary, MoE; Mr. Asif Shuja Khan, DG, Pak-EPA; Dr. Aurangzeb Khan, Chief (Environment), Planning Commission; Mr. Asadullah Faiz, Director, Pak-EPA; Mr. Muhammad Bashir Khan, Director, AJK EPA; Dr. Hussain Ahmad, Director, KP EPA; Mr. S. M. Yahya, Director, Sindh EPA; Dr. Muhammad Saleem, Director, Balochistan EPA; Mr. Nasir Ali Khan, Project Director, Pak-EPA; Mr. Rafiullah Khan, Deputy Director, Pak-EPA; Mr. Khadim Hussain, Assistant Director, GB EPA; Mr. Muhammad Farooq Alam, Research Officer (Air Pollution), Punjab EPD; Mr. Naseem Afzal Baz, Director (Environment), NHA;

Mr. Asif Jamil Akhter, Assistant Director, NHA; Ms. Shahzia Shahid, Environment Specialist, NHA; Mr. Kaleem Anwar Mir, Scientific Officer, GCISC; Dr. Mahmood Khawaja, Senior Advisor (Chemicals), SDPI; Mr. M. Ilyas Suri, Director, Indus Motors; Mr. Saad Murad Khan, Deputy Manager, Indus Motors; Ch. Usman Ali, Consultant, Metro Consulting; and Mr. Muhammad Qadiruddin, General Manager, Global Environmental Laboratories.

The team is particularly grateful to the governments of Australia, Finland, the Netherlands, and Norway for their support through several trust funds—particularly the TFSSD, the BNPP, and the Country Environmental Analysis Trust Fund—that financed the studies underpinning this book.

About the Authors

Ernesto Sánchez-Triana is Lead Environmental Specialist for the World Bank's Latin America and Caribbean Region. Prior to joining the Bank, he was a professor at the National University of Colombia and served as Director of Environmental Policy at Colombia's National Department of Planning. From 2006 to 2012, he worked for the Bank's South Asia Region, leading numerous operations including analytical work on "Policy Options for Air Quality Management in Pakistan"—the basis for this book. Dr. Sánchez-Triana holds MS and PhD degrees from Stanford University and has authored numerous publications on environmental economics, energy efficiency, environmental policy, poverty and social impact assessment, and green growth.

Santiago Enriquez is an international consultant with more than 15 years of experience in the design, implementation, and evaluation of policies relating to the environment, conservation, and climate change. He has developed analytical work for the World Bank, United States Agency for International Development, and the Inter-American Development Bank. From 1998 to 2002, Mr. Enriquez worked at the International Affairs Unit of Mexico's Ministry of Environment and Natural Resources. Mr. Enriquez holds a master's degree in public policy from the Harvard Kennedy School.

Javaid Afzal is a Senior Environment Specialist at the World Bank's Islamabad office. His responsibilities include moving the environment development agenda forward with client government agencies. He also task manages operations in water resources and the environment, and provides environmental safeguards support for the Bank's South Asia Region. Previously, he worked at a leading consulting company in Pakistan. Dr. Afzal holds a PhD in water resources management from Cranfield University, U.K., and master's and bachelor's degrees in agricultural engineering from the University of Agriculture, Faisalabad, Pakistan. Dr. Afzal has published in a number of peer-reviewed journals on the topics mentioned above.

Akiko Nakagawa is a Senior Environmental Specialist in the World Bank's South Asia Region Disaster Risk Management and Climate Change Unit at the Bank's headquarters, where she works on climate change mitigation and adaptation in the region. Before joining the Bank, she served in the Japanese government's Ministry of Environment as a climate change negotiator. Prior to joining the government, she worked at JBIC/OECF and managed a lending portfolio comprising projects such as watershed management, energy efficiency, afforestation, flood-resilient community development, and river embankment. She has a master's degree in urban planning (urban environment) from New York University.

Asif Shuja Khan is former Director General, Pakistan Environmental Protection Agency (Climate Change Division). He was the Director General of the Pakistan Environmental Protection Agency (Pak-EPA) since its inception in 1993 and until February 2014. He has more than 30 years of professional experience in environmental management and has worked with international agencies like the World Bank, Asian Development Bank, Japanese International Cooperation Agency, United Nations Environmental Programme, United Nations Development Programme, and United Nations International Children's Emergency Fund. Under his leadership, Pak-EPA established Air Quality Monitoring systems in Pakistan's major cities and adopted national ambient air quality standards and Euro emission standards for vehicles. Mr. Shuja holds a master's degree from the Stevens Institute of Technology, New Jersey, U.S.

Abbreviations

ALRI	acute lower respiratory infections
AOD	aerosol optical depth
AQM	air quality management
BCR	benefit-cost ratio
BLL	blood lead level
CB	chronic bronchitis
CBA	cost-benefit analysis
CCD	Climate Change Division
CNG	compressed natural gas
CO	carbon monoxide
CO_2	carbon dioxide
COI	cost of illness
DALY	disability-adjusted life year
DOC	diesel oxidation catalysts
DPF	diesel particulate filters
EPAs	Environmental Protection Agencies
EPI	environmental performance index
ERV	emergency room visit
ET	environmental tribunal
EU	European Union
GDI	gross domestic income
GDP	gross domestic product
GDP PPP	gross domestic product purchasing power parity
GHG	greenhouse gases
GoP	Government of Pakistan
HAD	hospital admission
HCV	human capital value

I&M	inspection and maintenance
IQ	intelligence quotient
JICA	Japanese International Cooperation Agency
LCV	light commercial vehicle
LPG	liquid petroleum gas
LRI	lower respiratory illness
MoA	Ministry of Agriculture
MoCC	Ministry of Climate Change
MoE	Ministry of Environment
MoEn	Ministry of Energy
MoF	Ministry of Finance
MoI	Ministry of Industry and Special Initiatives
MoPNR	Ministry of Petroleum and Natural Resources
NEP	National Environmental Policy
NEQS	National Environmental Quality Standards
NGO	nongovernmental organization
NO_2	nitrogen dioxide
NO_X	nitrogen oxides (nitric oxide and nitrogen dioxide)
O_3	ozone
OAP	outdoor air pollution
OC	organic carbon
OECD	Organisation for Economic Co-operation and Development
Pak-EPA	Pakistan Environmental Protection Agency
Pb	lead
PCA	principal component analysis
PCAP	Pakistan Clean Air Program
PEPA	Pakistan Environmental Protection Act of 1997
PEPC	Pakistan Environmental Protection Council
PM	particulate matter
PM_1	particulate matter of less than 1 micron
$PM_{2.5}$	particulate matter of less than 2.5 microns
PM_{10}	particulate matter of less than 10 microns
PMF	Positive Matrix Factorization
ppm	parts per million
PPP	polluter pays principle
RAD	restricted activity days

RS	respiratory symptoms
SMART	Self-Monitoring and Reporting Tool
SO_2	sulfur dioxide
SO_x	sulfur oxides
SPM	suspended particulate matter
SUPARCO	Pakistan Space and Upper Atmosphere Research Commission
TSP	total suspended particles
UNEP	United Nations Environment Programme
USEPA	United States Environmental Protection Agency
VOC	volatile organic compounds
VSL	value of a statistical life
WHO	World Health Organization
WTP	willingness to pay
YLL	years of life lost

Executive Summary

Policy Options to Address the Cost of Outdoor Air Pollution in Pakistan

Introduction

Examining options for improving Pakistan's air quality... This document examines options for Pakistan to address the significant costs that ever-worsening air pollution imposes upon its economy and populace. This book examines policy options to control Pakistan's outdoor air pollution, with the aim of assisting the Government of Pakistan (GoP) in designing and implementing reforms to improve Pakistan's ambient air quality regulatory framework. The book draws upon an analysis of air quality management (AQM) in Pakistan and is informed by international and regional best practices. The document proposes a prioritized menu of cost-effective interventions to improve Pakistan's AQM, particularly in urban areas, which face the most significant air quality problems.

Among world's worst urban air pollution... Pakistan's urban air pollution is among the most severe in the world, and it significantly damages human health and the economy. Pakistan is the most urbanized country in South Asia, and it is undergoing rapid motorization and increasing energy use. Air pollution, particularly in large urban centers, damages the populations' health and quality of life, and contributes to environmental degradation (Aziz 2006; Aziz and Bajwa 2004; Colbeck, Nasir, and Ali 2010b; Sánchez-Triana, Ahmed, and Awe 2007; World Bank 2006, 2007).[1] From 2007 to 2011, the reported levels of particulate matter (PM), sulfur dioxide (SO_2), and lead (Pb) were many times higher than the World Health Organization (WHO) air quality guidelines.[2,3] The number of premature deaths and illnesses caused by air pollution exceeds most other high-profile causes of public health problems that receive significantly more attention in Pakistan, including road accidents.[4]

Worsening expectations... Current trends, including industrialization and urbanization, suggest that air quality will worsen unless targeted interventions are adopted in the short, medium, and long term, and unless the institutional and technical capacity of organizations responsible for AQM are strengthened. This book advocates for allocating resources to AQM, because there is evidence that air quality is severely affecting millions of Pakistanis, and because experiences from around the world indicate that well-targeted interventions and an adequate institutional framework can significantly improve air quality.

More vehicles... The number of vehicles in Pakistan has jumped from approximately 2 million to 10.6 million over the last 20 years, an average annual growth rate in excess of 8.5%.[5] From 1991 to 2012, the number of motorcycles and scooters grew more than 450%, and motor cars, close to 650% (figure 1). The growth rate of mobile sources increased after 2003.

Industrial facilities... Industrial facilities, particularly those consuming fossil fuels, emit significant amounts of air pollutants. Emissions from large-scale facilities, such as cement, fertilizer, sugar, steel, and power plants—many of which use furnace oil that is high in sulfur content—are a major contributor to poor air quality (Ghauri, Lodhi, and Mansha 2007; Khan 2011). A wide range of small-scale to medium-scale industries, including brick kilns, steel re-rolling, steel recycling, and plastic molding, also contribute substantially to urban air pollution through their use of "waste" fuels, including old tires, paper, wood, and textile waste. Industrial emissions are further exacerbated by the widespread use of small diesel electric generators in commercial and residential areas in response to the electricity outages. Industrial emissions are associated with poor maintenance of boilers and generators (Colbeck, Nasir, and Ali 2010a; Ghauri 2010; Ilyas 2007; Khan 2011).[6]

Waste burning, dry weather, strong winds... Different nonpoint sources contribute to air pollution in Pakistan, including burning of solid wastes and sugarcane fields. More than 54,000 tons of solid waste are generated daily, most of which is either dumped in low-lying areas or burned. The burning of solid waste at low temperatures produces carbon monoxide (CO), PM, and volatile organic compounds (VOCs), including toxic and carcinogenic pollutants (Faiz 2011). Farmers in Pakistan burn cane fields to ease harvesting. During sugarcane harvesting, high concentrations of particulate matter of less than 10 microns (PM_{10}) are found in rural areas in Punjab and Sindh. Predominantly dry weather in arid conditions and strong winds also generate substantial dust in most parts of Sindh province and southern Punjab, elevating PM_{10} levels in the air. Due to high summer temperatures (40–50°C), fine dust is transported into the atmosphere with the rising hot air and forms "dust clouds" and haze over many cities of southern Punjab and upper Sindh. Dust storms are also generated from deserts (Thal, Cholistan, and Thar), particularly during the summer, and adversely affect air quality in the cities of Punjab and Sindh (Hussain, Mir, and Afzal 2005).

Figure 1 Motor Vehicles on Road (in thousands) in Pakistan, 1991–2012

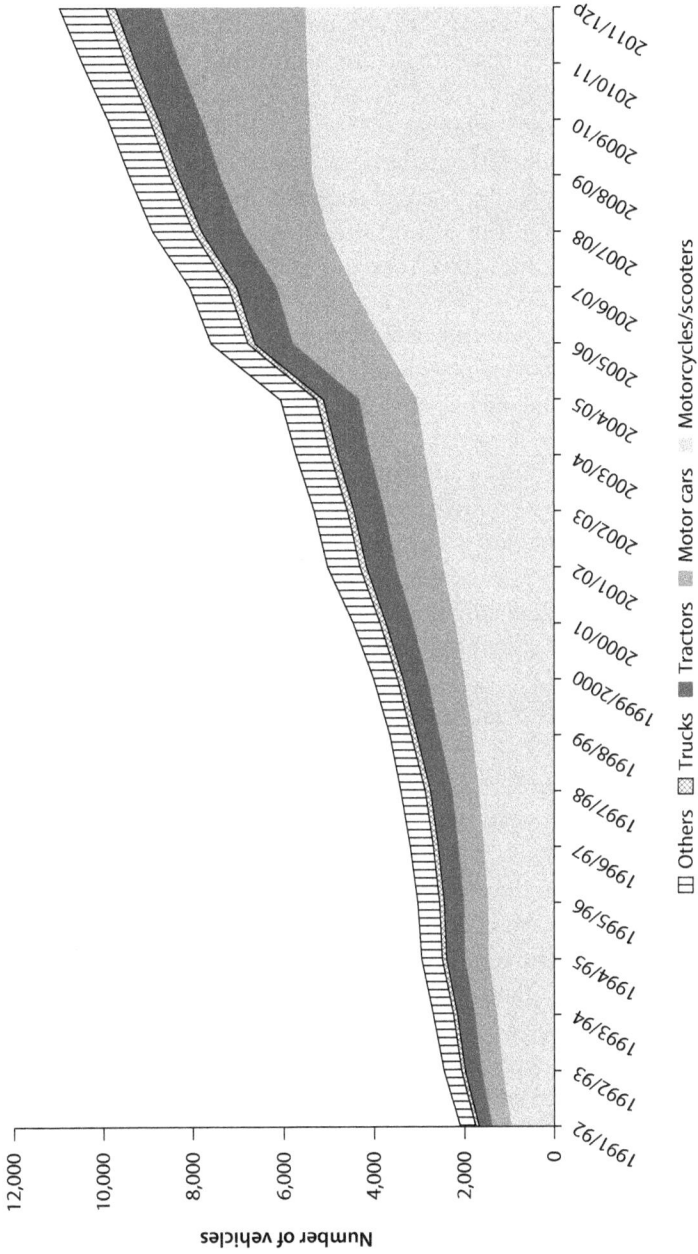

Legend: Others | Trucks | Tractors | Motor cars | Motorcycles/scooters

Y-axis: Number of vehicles

X-axis: 1991/92, 1992/93, 1993/94, 1994/95, 1995/96, 1996/97, 1997/98, 1998/99, 1999/2000, 2000/01, 2001/02, 2002/03, 2003/04, 2004/05, 2005/06, 2006/07, 2007/08, 2008/09, 2009/10, 2010/11, 2011/12p

Source: Ministry of Finance of Pakistan 2012.
Note: Data for 2011–12 are provisional, as indicated by the "p".

3

Particulate matter... Source apportionment analyses completed in Pakistan (particularly for Lahore) from 2006 to 2012 have found high concentrations of primary and secondary pollutants, particularly, fine PM. In Lahore, Stone and others (2010) reported the chemical characterization and source apportionment of fine and coarse PM.[7] The annual average concentration (±one standard deviation) of particulate matter of less than 2.5 microns ($PM_{2.5}$) was 194 ± 94 micrograms per cubic meter ($\mu g/m^3$). Crustal sources like dust dominated coarse aerosol, whereas carbonaceous aerosol dominated fine particles. While motor vehicle contributions were relatively consistent over the course of the yearlong study, biomass and coal sources demonstrated seasonal variability and peaked in the wintertime. Secondary organic aerosols' contributions also peaked in the wintertime, potentially enhanced by fog.

Industrial and mobile sources, soil, burning... In Lahore, dust sources were found to contribute on average 41% of PM_{10} mass and 14% of $PM_{2.5}$ mass on a monthly basis (von Schneidemesser and others 2010). Seasonally, concentrations were found to be lowest during the monsoon season (July–September). Principal component analysis (PCA) identified seven factors: three industrial sources, resuspended soil, mobile sources, and two regional secondary aerosol sources likely from coal and/or biomass burning. PM measured in Lahore was more than an order of magnitude greater than that measured in aerosols from the Long Beach/Los Angeles region and approximately four-fold greater than the activity of the Denver area PM (Shafer and others 2010).

Particulate pollution is 2 to 14 times USEPA levels... Elevated levels of $PM_{2.5}$ at Lahore ranged from 2 to 14 times higher than the prescribed United States Environmental Protection Agency (USEPA) limits (Lodhi and others 2009). Source apportionment was performed on short duration analysis results of November 2005 to March 2006, using a Positive Matrix Factorization (PMF) model. The results from the PMF model indicated that the major contributors to $PM_{2.5}$ in Lahore are soil/road dust, industrial emissions, vehicular emissions, and secondary aerosols (Lodhi and others 2009; Quraishi, Schauer, and Zhang 2009). Transboundary air pollutants also affect the city, particularly due to secondary aerosols during winter. The sulfate particles also facilitate haze/fog formation during calm and highly humid conditions, and thus reduce visibility and increase the incidence of respiratory diseases encountered in the city annually. Earlier studies (Hameed and others 2000) also reported similar events causing marked reductions in visibility, disrupted transportation, and increased injury and death.

Lahore pollution sources... Also in Lahore, from November 2005 to January 2006, several sources contributed to $PM_{2.5}$ concentrations, including diesel emissions (28%), biomass burning (15%), coal combustion (13%), secondary PM (30%), exhaust from two-stroke vehicles (8%), and industrial sources (6%) (Raja and others 2010). Diesel and two-stroke vehicle emissions accounted for much (36%) of the measured high $PM_{2.5}$. Although a large component of the carbonaceous aerosols in Lahore originated from fossil fuel combustion, a significant fraction was derived from biomass burning (Husain and others 2007). Finally, Zhang, Quraishi, and Schauer (2008), by using a molecular marker based Chemical Mass Balance (CMB) receptor model, showed that traffic pollution, including exhaust from gasoline- or diesel-powered vehicles, was the predominant source of PM_{10} carbonaceous aerosols. Gasoline-powered vehicles plus diesel exhausts contribute 47.5%, 88.3%, and 15.4% of measured inhalable particulate organic carbon, elemental carbon, and mass, respectively.

Karachi pollution sources... In Karachi, Mansha and others (2011) reported a characterization and source apportionment of ambient air particulate matter ($PM_{2.5}$). Source apportionment was performed on PM samples using a PMF model, indicating five major contributors: industrial emissions (53%), soil/road dust (16%), and others, including vehicular emissions (18%), sea salt originating from the Arabian Sea, and secondary aerosols.

Islamabad pollution sources... In Islamabad, the mean metal concentrations in the atmosphere are far higher than background and European urban sites, mainly due to anthropogenic activities (Shah and Shaheen 2009). Industrial metals like iron, zinc, manganese, and potassium showed viable correlations, while Pb is correlated with cadmium because of their common source. Principal component and cluster analyses revealed automobile emissions, industrial activities, combustion processes, and mineral dust as the major pollution sources of the atmospheric particles (Shah and Shaheen 2007b; Shah, Shaheen, and Nazir 2012). The comparison study presents high concentrations of airborne trace metal. The basic statistical data revealed quite divergent variations of the elements during the specific seasons (Shah and Shaheen 2010). Automobile emissions, windblown soil dust, excavation activities, biomass burning, and industrial and fugitive emissions were identified as major pollution sources in Islamabad's atmospheric aerosols. The comparative data showed that the concentrations of airborne trace elements in this area are mostly very high compared with other regions, thus posing a potential health hazard to the local population.

Data Availability and Methodology

Methodology for Air Quality Data Analysis

Network for air quality management (AQM)... The provincial Environmental Protection Agencies (EPAs) and the Pakistan EPA (Pak-EPA) are in charge of monitoring air pollution in Pakistan. From 2006 to 2009, the Japanese International Cooperation Agency (JICA) assisted the GoP in designing and installing an air quality monitoring network of measurement stations that included (a) fixed and mobile air monitoring stations in five major cities of Pakistan (Islamabad, Karachi, Lahore, Peshawar, and Quetta); (b) a data center; and (c) a central laboratory. The provincial EPAs managed and operated the monitoring units in the provinces. Consultants hired, trained, and paid by the

Japanese partners initially carried out the actual operation, and after the first year of operation, the EPAs were expected to assume all costs related to the monitoring work.

Inadequate operation and maintenance... Administrative and budget problems have led to inadequate operation and maintenance of the air quality monitoring network. Automated instruments were used in Lahore, Karachi, Quetta, Peshawar, and Islamabad, while manual, or manual and automated combinations, were used in other areas. The number and location of the monitors were criticized because they excluded "hot spots" where air quality was believed to be particularly poor. Various technical problems were reported related to the interrupted power supply and difficulties in maintaining the automated electronic instruments.

AQM network operations suspended... As of mid 2014, the EPAs had not assumed the operation and maintenance costs of the air quality monitoring network; consequently, the network suspended its operations. Furthermore, the data that were collected were neither analyzed nor disclosed, and the concentration of $PM_{2.5}$ was being monitored infrequently. As emphasized in this book, despite the installation of this air quality monitoring system, the reliability of air quality data are suboptimal in Pakistan. Furthermore, air quality data collected before 2004 are significantly less reliable than the data collected by this air quality monitoring network.

Availability of air quality data ranging from 5% to 96%... While there is a paucity of available air quality data for Pakistan, JICA's funded air quality monitoring network collected information on concentrations of PM, carbon monoxide, sulfur and nitrogen oxides, ozone, and other parameters in Pakistan's major cities.[8] However, daily average ambient air quality data are available for less than half of the days between July 26, 2007, and April 27, 2010. Table 1 shows the temporal (percent of days) coverage for all the air quality data, and indicates that $PM_{2.5}$ had the lowest coverage rate among key pollutants.

Despite its flaws, these data provide the best picture of Pakistan's air quality. Using these data, average values were calculated for all air quality pollutants.

Table 1 Data Coverage (%) for Air Quality Parameters for Five Cities in Pakistan

Data coverage (%)	$PM_{2.5}$	SO_2	NO_2	O_3	CO
Islamabad	42	31	78	87	96
Quetta	5	46	53	29	53
Karachi	10	23	29	24	24
Peshawar	17	35	66	28	60
Lahore	12	54	75	79	44
Average	17	38	60	49	55

Source: Dall'Osto 2012.
Note: $PM_{2.5}$ = particulate matter of less than 2.5 microns, SO_2 = sulfur dioxide, NO_2 = nitrogen dioxide, O_3 = ozone, CO = carbon monoxide.

PCA was performed using STATISTICA v4.2 software on a dataset composed of meteorological parameters, gaseous pollutants, and $PM_{2.5}$ mass. This methodology combines a factor analysis that results in the identification of potential pollution sources, indicates the seasonal evolution of the sources, and quantifies the annual mean contribution of each one.

Methodology for Economic Analysis

Broad spectrum of primary data sources... The environmental health and economic analysis presented in this book relies on primary data obtained from various ministries, agencies, and institutions in Pakistan, as well as from international development agencies. The analysis also uses several hundred reports and research studies from Pakistan and other countries. Quantification of health effects from environmental risk factors is grounded in commonly used methodologies that link health outcomes and exposure to pollution and to other health risk factors. The economic costs of these health effects are estimated using standard valuation techniques. The assessment of the benefits and costs of interventions to mitigate health effects and improve natural resource conditions is based on these same methodologies and valuation techniques, as well as on international evidence of intervention effectiveness, and to the extent available, on data regarding the costs of interventions in Pakistan.

Methodology for Institutional Analysis

Institutional foundations, weaknesses, funding... Pakistan has institutional foundations that could support the development of priority actions targeting the country's severe air pollution problem. However, a number of obstacles have hampered the development of adequate responses, including acute institutional weaknesses and lack of funding. The institutional analysis included in this book assessed the legal mandates of environmental organizations at the federal and provincial levels, particularly after the 18th Constitutional Amendment. Based on those mandates, the analysis evaluates the adequacy of human resources to carry out key technical, management, and support functions, as well as the availability of physical capital needed to perform the assigned functions. The institutional analysis includes a review of the existing formal rules (for example, laws, policies, and standards), informal rules (for example, the capacity of powerful groups to reverse the adoption of pollution charges, and the pervasive tolerance of violators of environmental regulations), and their enforcement mechanisms. In addition, the analysis looks at the role that the courts have played and could play in the future when environmental organizations fail to fulfill their AQM responsibilities.

Analysis of Air Quality Data

Concentrations of Air Pollutants

Very high concentrations of fine particulate matter... An analysis of the available data from 2007 to 2010 shows very high concentrations of fine particulate matter ($PM_{2.5}$)—measured in micrograms per cubic meter ($\mu g/m^3$)—in Lahore (143 $\mu g/m^3$), Karachi (88 $\mu g/m^3$), Peshawar (71 $\mu g/m^3$), Islamabad (61 $\mu g/m^3$), and Quetta (49 $\mu g/m^3$). Most likely, the high value concentrations reported in this analysis would have been even higher if the monitoring instruments had been working all the time. Particulate matter of less than 1 micron (PM_1) and PM_{10} measurements were not available. Low data coverage (average of 17% for the five cities) partially affected the $PM_{2.5}$ measurements.

Lahore had highest sulfur dioxide concentrations... The analysis of the 2007–10 time series on SO_2 confirmed that Lahore was the city with the highest concentrations (74 ± 48 $\mu g/m^3$), with maximum daily values of 309 $\mu g/m^3$. Other cities presented very high values of SO_2: Quetta (54 ± 26 $\mu g/m^3$), Peshawar (39 ± 34 $\mu g/m^3$), and Karachi (34 ± 34 $\mu g/m^3$). Overall, SO_2 values were found to be increasing over the course of the study period (2007–10).

Pakistan's nitrogen dioxide concentrations slightly above WHO guideline... The annual nitrogen dioxide (NO_2) concentrations derived from the 48-hour data revealed that the current levels in the country are slightly higher than the WHO air quality guideline value of 40 $\mu g/m^3$, with the highest concentrations in Peshawar (52 ± 21 $\mu g/m^3$), Islamabad (49 ± 28 $\mu g/m^3$), Lahore (49 ± 25 $\mu g/m^3$), and Karachi (46 ± 15 $\mu g/m^3$). Concentrations were somewhat lower in Quetta (37 ± 15 $\mu g/m^3$). Results from an analysis of data from 2007 to 2011 show that concentrations of ozone (O_3) and CO were well within the WHO guidelines.

Temporal Trends and Source Apportionment of Air Pollutants
Seasonal impacts on air pollutants... An analysis of temporal trends of air quality pollutants for Islamabad, Karachi, Peshawar, and Lahore shows higher CO, NO_2, SO_2, and $PM_{2.5}$ concentrations during winter periods, whereas O_3 shows the opposite trend. Since sunlight and heat drive ozone formation, warm sunny days usually have more ozone than cool or cloudy days. All pollutants present similar wind roses. Thermal inversions, which happen across much of Pakistan from December to March, lower the mixing height and result in high pollutant concentrations, especially under stable atmospheric conditions.

Suspended PM contributes to the formation of ground fog that prevails over much of the Indo-Gangetic Plains during the winter months. In addition, sunny and stable weather conditions lead to high concentrations of pollutants in the atmosphere (Sami, Waseem, and Akbar 2006). Due to high temperatures in summer (40–50°C), fine dust rises with the hot air and forms "dust clouds" and haze over many cities of southern Punjab and upper Sindh. Dust storms generated from deserts (Thal, Cholistan, and Thar), particularly during the summer, adversely affect visibility in the cities of Punjab and Sindh (figure 2).

Source Apportionment of Air Quality Pollutants
Road traffic a main source... A statistical analysis shows that the strongest correlations (expressed as R^2) among key parameters are found between $PM_{2.5}$ and

Figure 2   Temporal Trend for PM$_{2.5}$, SO$_2$, NO$_2$, O$_3$, and CO for Five Pakistan Cities: Islamabad (ISL), Quetta (QUE), Lahore (LAH), Karachi (KAR), and Peshawar (PES)

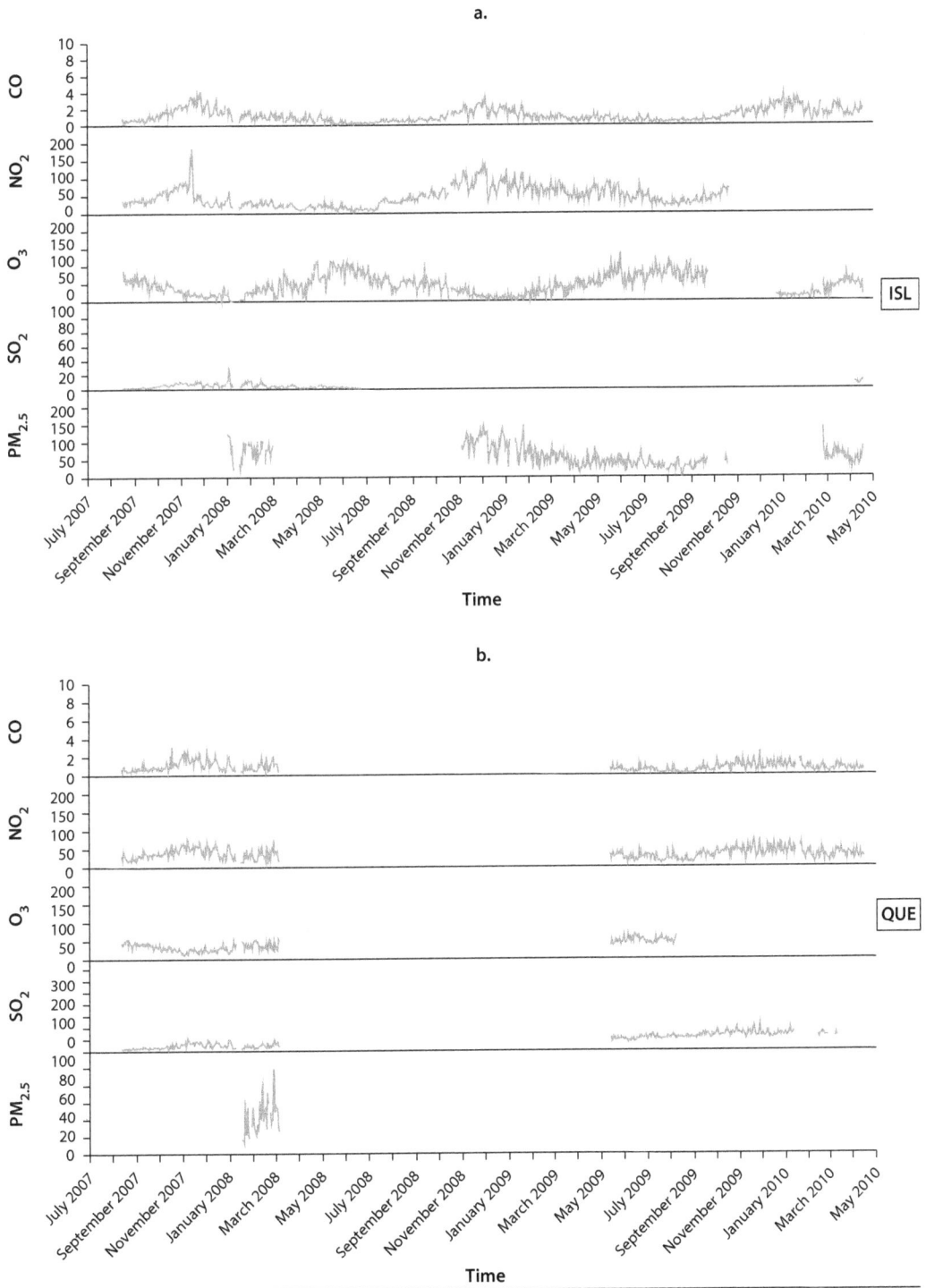

a.

b.

figure continues next page

Figure 2 Temporal Trend for PM$_{2.5}$, SO$_2$, NO$_2$, O$_3$, and CO for Five Pakistan Cities: Islamabad (ISL), Quetta (QUE), Lahore (LAH), Karachi (KAR), and Peshawar (PES) *(continued)*

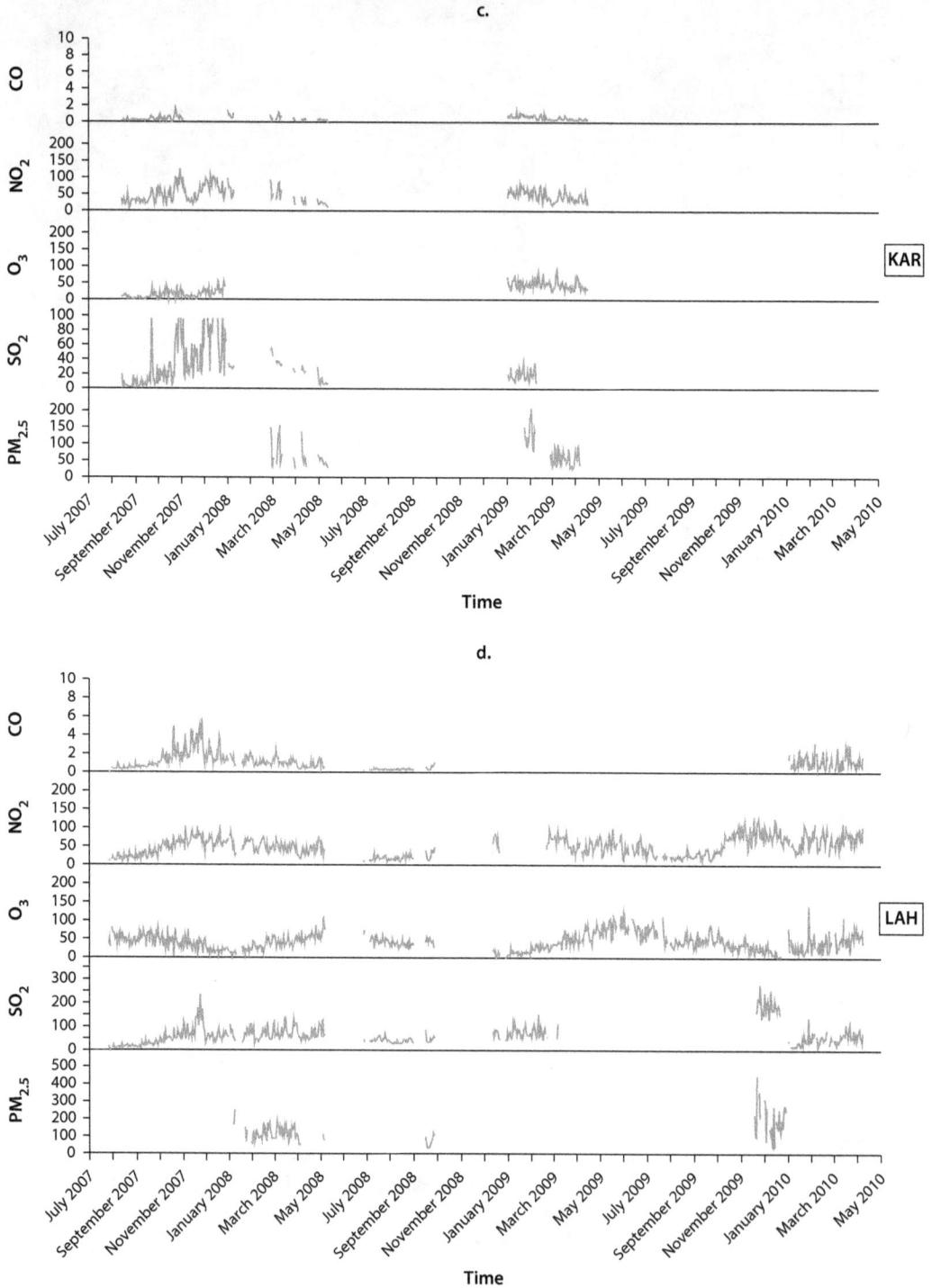

c.

d.

figure continues next page

Figure 2   Temporal Trend for PM$_{2.5}$, SO$_2$, NO$_2$, O$_3$, and CO for Five Pakistan Cities: Islamabad (ISL), Quetta (QUE), Lahore (LAH), Karachi (KAR), and Peshawar (PES) *(continued)*

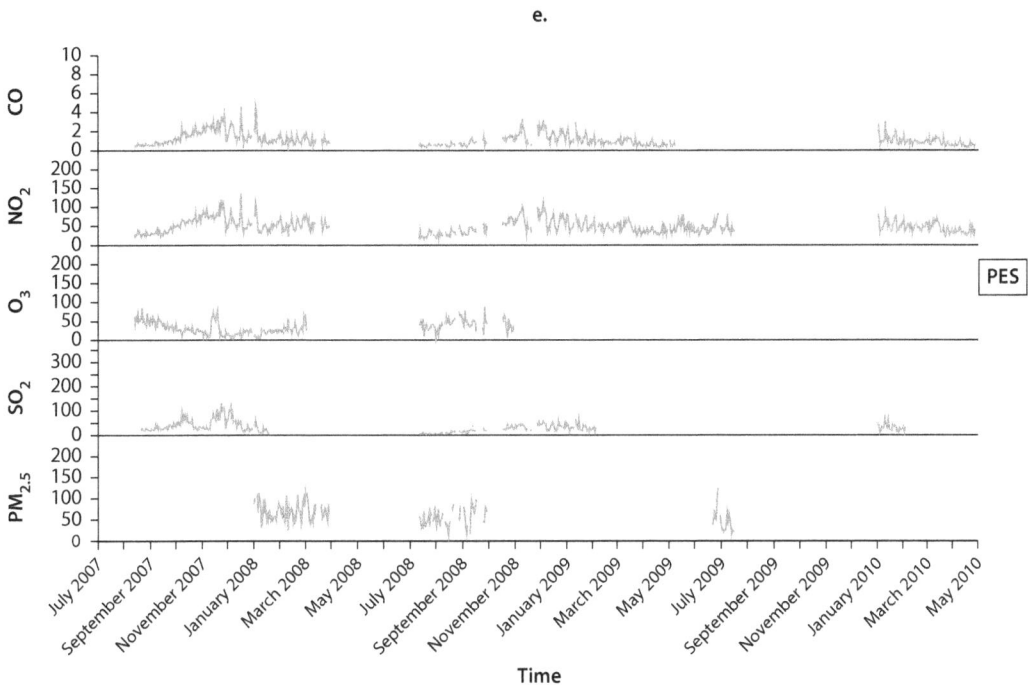

e.

Source: Dall'Osto 2012.
Note: PM$_{2.5}$ = particulate matter of less than 2.5 microns, SO$_2$ = sulfur dioxide, NO$_2$ = nitrogen dioxide, O$_3$ = ozone, CO = carbon monoxide.

CO, implying road traffic is a main source of fine PM in Pakistan (figure 3). Since PM$_{2.5}$ correlates better with CO than with SO$_2$ and NO$_2$, it is possible that fresh direct traffic emissions are important contributions to the fine particulate mass levels. Other sources—including industries and natural dust or sea salt—may also contribute, but our analysis suggests that direct traffic emissions are more related to high concentrations of ambient aerosols. The low correlations obtained at Lahore may be due to the very high levels of PM (which may be associated with natural factors, such as dust and emissions from industrial and agricultural sources).

Human-caused and weather-caused... When considering factor analysis, the "primary" factor includes traffic (NO$_2$, CO) and industrial (SO$_2$) gases, as well as PM$_{2.5}$. The "primary" factor is mainly associated with primary anthropogenic emissions, and it represents the major component for the cities of Islamabad, Karachi, and Peshawar (33–43%). The "secondary" factor is seen mainly during summer and is associated with ozone and high temperatures. It was the second main factor found (22–33%), and it is associated more with summer regional pollution events (table 2).

Cleaning Pakistan's Air • http://dx.doi.org/10.1596/978-1-4648-0235-5

Figure 3 Correlations between $PM_{2.5}$, SO_2, NO_2, O_3, and CO for Islamabad, Quetta, Karachi, Peshawar, and Lahore

City	R^2	SO_2	NO_2	O_3	CO
Islamabad	$PM_{2.5}$	0.35	0.45	−0.37	0.7
	SO_2		0.5	−0.25	0.5
	NO_2			−0.2	0.4
	O_3				−0.4
Quetta	$PM_{2.5}$	0.55	0.75	−0.4	0.85
	SO_2		0.7	−0.45	0.7
	NO_2			−0.55	0.85
	O_3				−0.5
Karachi	$PM_{2.5}$	0.2	0.4	0	0.45
	SO_2		0.55	0	0.1
	NO_2			0	0.7
	O_3				0
Peshawar	$PM_{2.5}$	0.2	0.5	0	0.6
	SO_2		0.4	0	0.4
	NO_2			0	0.8
	O_3				0
Lahore	$PM_{2.5}$	0	0	0	0
	SO_2		0.6	0	0.6
	NO_2			0	0.6
	O_3				n.a.

Source: Dall'Osto 2012.
Note: $PM_{2.5}$ = particulate matter of less than 2.5 microns, SO_2 = sulfur dioxide, NO_2 = nitrogen dioxide, O_3 = ozone, CO = carbon monoxide.

Human sources are major, natural sources minor... The analysis found that $PM_{2.5}$ is strongly correlated with CO and NO_2, indicating the importance of road traffic as a source, especially in winter months. By contrast, in the summer months, with higher wind speed, the influence of resuspended surface dusts and soils, and of secondary PM, may play a bigger role. However, the strong correlation between road traffic markers (such as CO) and $PM_{2.5}$ suggests that anthropogenic pollution is a major source of fine PM and that natural sources are minor ones.

Health Damages Associated with Air Pollution

Pollution significantly exceeds national and WHO standards... Limited available evidence indicates that concentrations of $PM_{2.5}$ in Islamabad, Karachi, Lahore, Peshawar, Rawalpindi, and Quetta are significantly above the National Environmental Quality Standards (NEQS) that came into force in July 2010, the stricter standards slated to come into force in January 2013, and the WHO guidelines (table 3). As examples of international best practices, large

Table 2 Principal Component Analysis: Results for Islamabad, Lahore, Karachi, and Peshawar

City	Primary	Secondary	Visibility
	PCA factors		
Islamabad			
$PM_{2.5}$	0.9	0.1	0.0
SO_2	0.9	0.2	0.0
NO_2	0.9	0.2	0.0
O_3	0.1	0.8	0.2
CO	0.9	0.0	0.1
Variance explained (%)	33	27	10
Karachi			
SO_2	0.9	0.1	0.3
NO_2	0.9	0.2	0.2
O_3	0.1	0.9	0.1
CO	0.7	0.5	0.1
Variance explained (%)	35	22	14
Peshawar			
$PM_{2.5}$	0.7	0.1	0.3
SO_2	0.9	0.3	0.2
NO_2	0.9	0.3	0.1
O_3	0.5	0.6	0.2
CO	0.9	0.0	0.1
Variance explained (%)	43	23	11
Lahore			
$PM_{2.5}$	0.6	0.3	0.3
SO_2	0.8	0.2	0.2
NO_2	0.8	0.5	0.2
O_3	0.2	0.8	0.1
CO	0.8	0.4	0.1
Variance explained (%)	22	33	11

Source: Dall'Osto 2012.
Note: PCA = principal component analysis, $PM_{2.5}$ = particulate matter of less than 2.5 microns, SO_2 = sulfur dioxide, NO_2 = nitrogen dioxide, O_3 = ozone, CO = carbon monoxide.

Table 3 Concentration Levels (Daily Average) of Suspended $PM_{2.5}$ in Pakistan
$\mu g/m^3$

Authors	Area	Year	Islamabad	Lahore	Karachi	Peshawar	Quetta
Pak-EPA 2007	6 sites	2005	–	–	–	–	104–222
Husain and others 2007	–	2005–06	–	53–476	–	–	–
Ghauri 2010[a]	Mobile stations	2007	43.7	–	–	–	–
	Fixed stations	2007	47.2	74.6	71.7	185.5	206.4

Sources: Colbeck, Nasir, and Ali 2010a, 2010b; Ghauri 2010 from data collected at stations installed by the Japanese International Cooperation Agency.
Note: — = not available. $PM_{2.5}$ (particulate matter of less than 2.5 microns) World Health Organization Guidelines = 10 $\mu g/m^3$; European Union Ambient Air Quality Standards = 25 $\mu g/m^3$; United States Ambient Air Quality Standards = 25 $\mu g/m^3$; Pakistan 2010 National Environmental Quality Standards = 25 $\mu g/m^3$.
a. The results of Ghauri (2010) are part of a World Bank-financed study reporting data obtained through interviews and primary data collected from monitoring stations in Pakistan.

metropolises, such as Mexico City, Santiago, and Bangkok, have successfully reduced their ambient concentrations of $PM_{2.5}$ to a level that is even lower than the level of small Pakistani cities, such as Gujranwala (Lodhi and others 2009).

Percent breakdown of pollution sources... Annual average $PM_{2.5}$ ambient air concentrations in Karachi are estimated at 88 μg/m^3 (figure 4). The analysis of primary data from 2006 to 2009 yielded the following estimates:

- 24–28% of ambient $PM_{2.5}$ concentrations in Karachi is from road vehicles
- 23–24% is from area-wide sources, including natural dust, resuspended road dust due to poor street cleaning, construction dust, agricultural residue burning, and salt particles from the sea
- 19–20% is from industry
- 12–13% is secondary particulates (sulfates and nitrates) formed in the atmosphere from sulfur dioxide and nitrogen oxides from combustion of fossil fuels
- 8–14% is from burning of solid waste in the city
- 4–5% is from domestic use of wood/biomass
- 2.3–2.8% is from oil and natural gas consumption by power plants and the domestic/public/commercial sectors.

Around two-thirds of PM emissions from road vehicles are estimated to be from diesel trucks, diesel buses and minibuses, and light-duty diesel vehicles. About one-third, however, appears to be from motorcycles and rickshaws, which almost exclusively have highly polluting two-stroke engines. A major source of PM emissions from industry corresponds to ferrous metal sources (steel mills, foundries, and scrap smelters).

Figure 4 Estimate of Annual Average PM$_{2.5}$ Ambient Air Concentrations, Karachi, 2006–09

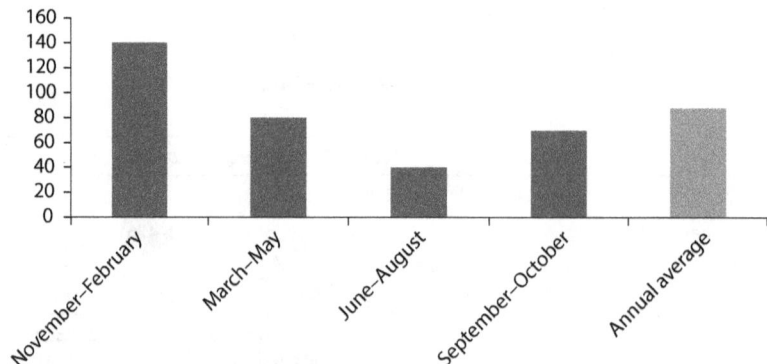

Sources: Sánchez-Triana and others 2014. Based on Alam, Trautmann, and Blaschke 2011; Ghauri 2008; Mansha and others 2011; and Sindh EPA 2010.
Note: PM$_{2.5}$ = particulate matter of less than 2.5 microns.

These $PM_{2.5}$ concentrations are estimated to cause over 9,000 premature deaths each year, representing 20% of acute lower respiratory infections (ALRI) mortality among children under five years of age, 24% of cardiopulmonary mortality, and 41% of lung cancer mortality among adults 30 or more years of age in these cities. About 12% of the deaths are among children under five years of age and 88% are among adults. Nearly 80% of the deaths are in Karachi.[9]

Annually: 22,600 deaths, 80,000 hospital admissions, 5 million childhood illnesses... By 2005, more than 22,600 deaths per year were directly or indirectly attributable to ambient air pollution in Pakistan, of which more than 800 are children under five years of age (World Bank 2006, 2008a). Outdoor air pollution alone caused more than 80,000 hospital admissions per year, nearly 8,000 cases of chronic bronchitis, and almost 5 million cases of lower respiratory cases among children under five. To put these numbers in perspective, the harm done by air pollution exceeds most other high-profile causes of mortality and morbidity that receive significantly more attention in Pakistan, including road accidents, which resulted in over 5,500 reported fatalities and nearly 13,000 non-fatal injuries in 2007 (WHO 2009).[10]

Air Quality Regulatory Framework

Background of Pakistan's AQM system... The framework for Pakistan's AQM system dates back to 1993, when the NEQS were developed under the 1983 Environmental Protection Ordinance. Consultations with major stakeholders were initiated in April 1996. In December 1999, the Pakistan Environmental Protection Council (PEPC) approved a revised version of the NEQS, and they became effective in August 2000. The review was justified by the PEPC because some of the original parameters were more stringent than parameters for other countries in South Asia.[11]

Drafting of NEQS for ambient air... In 2010, Pak-EPA drafted NEQS for ambient air. The NEQS for ambient air cover several major pollutants: (a) SO_2, (b) nitrogen oxide (NO_x), (c) O_3, (d) suspended particulate matter (SPM), (e) $PM_{2.5}$, (f) Pb, and (g) CO.[12] As required by law, prior to submitting the standards for PEPC's review and approval, Pak-EPA published the draft NEQS on its website and requested comments from the public. PEPC approved both standards in a meeting held on March 29, 2010, and the official notifications in the *Gazette of Pakistan* were made on November 26, 2010.[13]

Legislative cornerstone... The cornerstone of environmental legislation is the Pakistan Environmental Protection Act (PEPA), enacted on December 6, 1997. PEPA provides a comprehensive framework for regulating environmental protection, including air pollution. PEPA established the general conditions, prohibitions, penalties, and enforcement to prevent and control pollution, and to promote sustainable development. PEPA delineated the responsibilities of the PEPC, Pak-EPA, and provincial EPAs.

Pollution charges... Pakistan pollution charges established by PEPA 1997 provide an opportunity to implement the polluter pays principle (PPP). Following PEPA's dispositions, Pak-EPA—in consultation with industry, industry associations, nongovernmental organizations (NGOs), and public sector stakeholders—agreed on pollution charges for industrial sources. The Pollution Charge Rules 2001 included detailed formulas for reporting and paying pollution charges, defined parameters for pollution charges, and included clauses for escalation. After giving notification of these comprehensive rules, Pak-EPA did not implement the Pollution Charge Rules 2001, due to resistance from powerful groups (Khan 2011).

Pakistan Clean Air Program (PCAP)... In 2005, the Government of Pakistan elaborated the Pakistan Clean Air Program (PCAP), which contains a list of interventions for improving air quality. The PCAP includes interventions bearing on (a) vehicular emissions, (b) industrial emissions, (c) burning of solid waste, and (d) natural dust. PCAP's main objective is to reduce the health and economic impacts of air pollution by implementing a number of short-term and long-term measures that require action at all levels of government.

Ministry of Environment... The Ministry of Environment (MoE) was established as a full-fledged ministry in 2002. The MoE was second in the hierarchy of environmental institutions after PEPC, and its formal mandate comprised the design and implementation of national environmental policies, plans, and programs, including environmental planning, pollution control and prevention, and ecology. The 18th Constitutional Amendment, adopted in 2010, devolved all these responsibilities to the provinces. In parallel with the 18th Constitutional Amendment, the Ministry of National Disaster Management absorbed the MoE in October 2011. Since the GoP elevated the issue of climate change to cabinet level, the ministry was given an additional role to address climate change and was thus transformed into the Ministry of Climate Change (MoCC) in April 2012. However, in June 2013, the MoCC was downgraded from a ministry to a division, and more than 60% of its budget was cut.

Pak-EPA... The Pak-EPA is the federal agency, dependent on the Climate Change Division (CCD), responsible for implementing the PEPA in the national territory. Pak-EPA has a broad range of functions, including the administration and implementation of PEPA, and the associated rules and regulations.

No specific unit responsible for AQM... Neither the organizational structure of the CCD, nor that of Pak-EPA, has a specific unit or department that is responsible for AQM. In order to address priority air pollution problems,

Pak-EPA has proposed establishing units specialized in AQM at the national and provincial levels. The Pak-EPA would take over responsibilities for coordinating, designing, and implementing air quality policies. Technical cells, EPA at the national level, and technical cells at provincial agencies would be responsible for monitoring ambient air quality, and mobile, stationary, and diffuse emissions. Additionally, other technical cells would be responsible for regulatory enforcement and compliance. These units would be responsible for the effective planning and implementation of air quality standards in the territory where the national and provincial environmental organizations are directly competent. These units would also play a key role in coordinating works of the same nature carried out by all the agencies that comprise the institutional AQM network.

Provincial EPAs' mandate... After the delegation of functions from the federal government, the mandate of the provincial EPAs has become more comprehensive. Provincial EPAs have authority to handle the environmental management of their respective provinces. Their mandate includes implementing rules and regulations prepared under PEPA 1997 and additional legislation, per the needs of each province; designing, implementing, and enforcing environmental standards and regulations; and developing provincial systems for the implementation of pollution charges, among other responsibilities.

Specific AQM units essential... In response to allocated functions, provincial EPAs are strengthening their organizational structures. Provincial EPAs need capacity development in AQM, including in designing and implementing provincial policies; establishing, operating, and maintaining air quality monitoring systems within their operational area; and enforcing air quality regulations. The establishment of specific units within the organization of provincial agencies that exclusively deal with air pollution control has become essential for the design, implementation, and enforcement of air quality regulations.

Intergovernmental and intersectoral coordination needed... The physical boundaries of air pollution rarely coincide with those of existing administrative or political jurisdictions (districts, municipalities, and provinces). In addition, several sectoral ministries are important players in the design and implementation of AQM policies. As a result, the need for intergovernmental and intersectoral coordination emerges, nationally and internationally, vertically and horizontally. However, no formal mechanisms have existed (to June 2013) for agencies involved in environmental management to participate in a consultative process with other provincial or sectoral agencies for setting priorities, for the design and implementation of interventions, or for the monitoring and evaluation of effectiveness. Intersectoral coordination for the oversight of crosscutting issues is also nonexistent. Some attempts have been made to establish focal points within other non-environment ministries, but interactions among these focal points have not yet been institutionalized.

Absence of apex agency can lead to excessive variability... Most countries in the world currently have an apex central environmental ministry or agency with a number of technical and action-oriented agencies designating and implementing public policies and enforcing regulations. However, without proper coordination, decentralization eventually leads to significant differences in environmental quality across regions. In Pakistan, the 18th Constitutional Amendment has empowered provincial EPAs to take care of most of the environmental issues in the provinces, while Pak-EPA's main responsibilities have been limited to assisting provincial governments in the formulation of rules and regulations under PEPA 1997.

The judiciary... In Pakistan, the judiciary has played an increasingly important role in the enforcement of environmental laws. When regulatory avenues for environmental enforcement fail, the judicial system is often the only other recourse for resolving environmental conflicts. The Supreme Court of Pakistan has considered cases regarding the degradation of the environment and has concluded that the right to a clean environment is a fundamental right of all citizens of Pakistan. The High Courts in the provinces have also intervened and rendered decisions affecting future environmental management. As an example, the Lahore High Court appointed the Lahore Clean Air Commission to develop and submit a report on feasible and specific solutions and measures for monitoring, controlling, and improving vehicular air pollution in the city of Lahore. While Pakistan's Supreme Court has considered several environmental degradation and protection cases, further efforts are needed to bring the Court's involvement in AQM to fruition.

Air Quality Management and Climate Change

Vulnerability to climate change... Pakistan is one of the most vulnerable countries to climate change, as evidenced by the 2010 floods. In addition, emissions of greenhouse gases (GHG) from Pakistan are significantly lower than those from developed countries. However, Pakistan produces more carbon dioxide (CO_2) emissions per gross domestic product than any of its neighbor countries in the South Asia region, including the major contributor, India (World Bank 2010). Pakistan's overall trend indicates a steady increase in emissions per unit of economic output over the last decades (figure 5). Pakistan can take advantage of opportunities to both reduce its carbon intensity and improve local air quality.

Fossil fuel combustion accounts for more than 90% of total CO_2 emissions, and for 40% of overall GHG emissions in Pakistan. The second main contributor to overall GHG emissions is the agricultural sector, which is responsible for most of the methane and nitrous oxide emissions. Within the emissions produced by fuel combustion, the transport sector accounts for 23% of all CO_2 emissions, of which the road sector represents the largest part (97%) due to its continued dependence upon CO_2 intensive fuels. Only countries with a very small industrial and energy installed capacity show a higher contribution from the road

Figure 5 GHG Emissions in South Asia

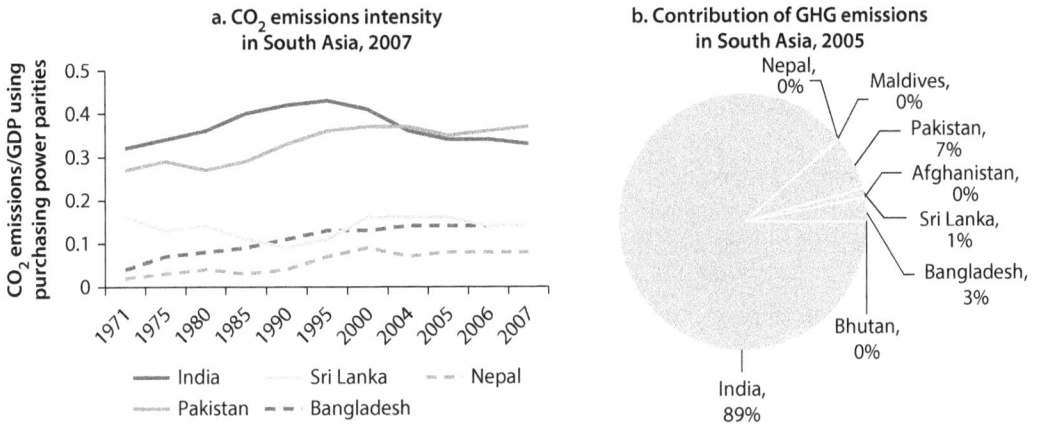

a. CO_2 emissions intensity in South Asia, 2007

b. Contribution of GHG emissions in South Asia, 2005

Nepal, 0%
Maldives, 0%
Pakistan, 7%
Afghanistan, 0%
Sri Lanka, 1%
Bangladesh, 3%
Bhutan, 0%
India, 89%

Sources: IEA 2009; World Bank 2010.
Note: GDP = gross domestic product; GHG = greenhouse gases.

sector to GHG emissions. Based on the use of commercial energy, industry seems to account for a relatively small share of total GHG emissions in Pakistan.

Co-benefits... Energy, transport, and agriculture are the sectors in Pakistan with significant climate change mitigation and air pollution control co-benefits. Energy-sector policies that generate environmental co-benefits include removing fuel subsidies, fuel switching, improving energy efficiency of plants, and renewable energy uptake.

Natural gas... Fuel switching from coal, fuel oil, or diesel to natural gas has the potential to bring about substantial improvements in local air quality and mitigate GHG emissions. Natural gas has the potential to produce less CO_2 per kilometer of travel than most other fossil fuels. It has been estimated that natural gas vehicles can reduce GHG emissions by as much as 20–25% over gasoline vehicles (IANGV 2003, 5). The growth in the use of compressed natural gas (CNG) in Pakistan has also provided economic benefits through employment generation.

Tailoring transport policies... Transport policies can be tailored in Pakistan to produce co-benefits in the form of reductions in local air pollutants, with the associated health benefits and reductions in congestion, noise, and accidents. In Pakistan, transport policies with significant co-benefits include fostering the use of railways or modernizing the trucking sector. Other climate policies in the transportation sector include improving the efficiency of motorized vehicles; promoting mass transit and policies to reduce congestion on roads, highways, and urban metropolitan centers; and promotion of non-motorized transport (Sánchez-Triana and others 2013).

Climate policies... Climate policies in the agricultural sector include improved soil management practices. Due to high summer temperatures (40–50°C), fine dust is transported into the atmosphere with the rising hot air and forms "dust clouds" and haze over many cities of southern Punjab and upper Sindh. Dust storms are also generated from deserts (Thal, Cholistan, and Thar), particularly during the summer, and adversely affect air quality in the cities of Punjab and Sindh. Improved soil management practices increase fertility and soil quality, while at the same time enhancing adaptation to drought by improving the soil's water content and resource conservation. Soil and water management enhances soil carbon sequestration (through increased organic matter residues returned to soil) and reduces emissions from land use, land-use change, and intensive agriculture practices. These actions can play an important role in voluntary carbon markets by promoting the creation of large carbon sinks and stocks (Akbar and Hamilton 2010).

Policy Options for Abating Urban Air Pollution from Mobile Sources

Managing emissions from mobile sources... Possible interventions to control and reduce PM emissions from mobile sources include (a) reducing sulfur in diesel and fuel oil, (b) retrofitting in-use diesel vehicles with PM emission-control technology, (c) converting diesel-fueled minibuses and vans to CNG,[14] (d) controlling PM emissions from motorcycles, and (e) converting three-wheelers (rickshaws) to CNG. These interventions would not only reduce PM emissions, but low-sulfur fuels would also reduce secondary particulates by reducing sulfur dioxide emissions. Other potential interventions that should be assessed include curtailing burning of solid waste in the city (and using the informal recycling industry), controlling PM emissions from ferrous metal sources and other industrial sources, improving street cleaning, and controlling construction dust.

Reducing sulfur content... Sulfur in diesel is being reduced to 500 parts per million (ppm) in Pakistan since 2012, but no confirmed timetable has been established for 50-ppm sulfur diesel. The analysis estimates that the health benefits of using 500-ppm sulfur content in diesel in road transport amount to at least US$2.3–3.5 per barrel of diesel for light diesel vehicles and large diesel buses and trucks used primarily within Karachi. Lowering the sulfur content further to 50 ppm would provide additional health benefits of US$3.0–4.6 per barrel.[15] This compares to an approximate cost of US$1.5–2.5 per barrel for lowering the sulfur content to 500 ppm and US$2–3 per barrel for lowering sulfur from 500 ppm to 50 ppm. Thus, the midpoint estimated health benefits per dollar spent (that is, benefit-cost ratio) on cleaner diesel are in the range of about US$1.1–1.2 for light-duty diesel vehicles, and US$1.7–1.8 for large buses and trucks, for both 500-ppm and 50-ppm diesel (figure 6).

Health benefits of reducing sulfur... Fuel oil in Pakistan generally has a sulfur content that averages around 3%, but some fuel oil with 1% sulfur is being imported. PM emission rates from combustion of fuel oil are greatly influenced by the sulfur content. Reducing sulfur from 3% to 1% is estimated to have health benefits of US$35–47 per ton of fuel oil. The additional cost of

Figure 6 Benefit-Cost Ratios of Low-Sulfur Fuels in Karachi

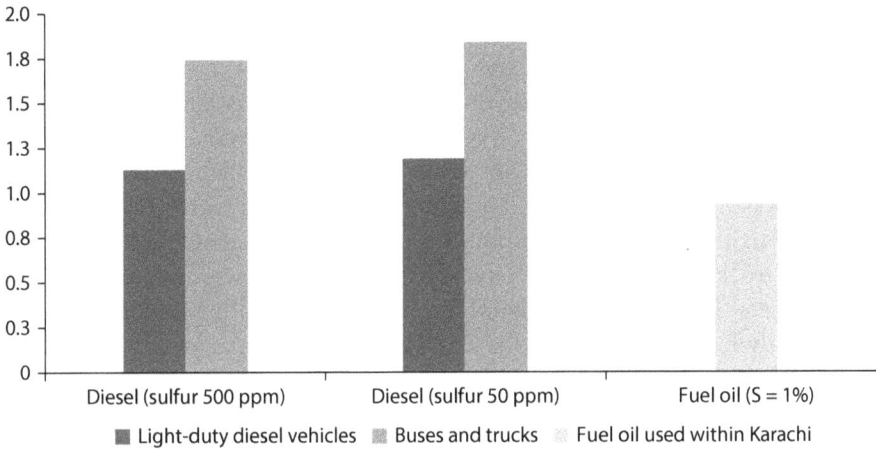

Source: Sánchez-Triana and others 2014.
Note: Midpoint estimate of incremental cost of low-sulfur fuels. S is the sulfur content in fuel oil.

low-sulfur fuel oil in the international markets fluctuates and has recently been around US$50 per ton. Thus, use of low-sulfur fuel oil should be targeted at users within urban centers. Additional health benefits of low-sulfur fuel oil include reduced sulfur dioxide emissions and, thus, lower secondary particulates formation.

Catalysts and filters... More stringent PM emission standards and control options can be implemented for diesel vehicles once low-sulfur diesel is available. Euro standards can be mandated on new diesel vehicles (and second-hand imports), and PM control technologies can effectively be installed on in-use diesel vehicles, such as diesel oxidation catalysts (DOCs) and diesel particulate filters (DPFs).

DOCs require a maximum of 500-ppm sulfur in diesel and DPFs require a maximum of 50 ppm to function effectively. A DOC generally reduces PM emissions by 20–30%, while a DPF reduces PM by more than 85%. DOCs had already been installed on over 50 million diesel passenger vehicles and more than 1.5 million buses and trucks worldwide five or six years ago.

Practices in other countries... All new on-road diesel vehicles in the United States and Canada are equipped with a high-efficiency DPF, while all new diesel cars and vans in the European Union have been equipped with DPF since 2009. Worldwide, over 200,000 heavy-duty vehicles had already been retrofitted with DPF about six years ago. DOCs and DPFs have also been used for retrofitting buses and trucks in many countries and locations, both on a wide scale and in demonstration projects.

Potential candidates for retrofitting with a DOC, or with a DPF when 50-ppm sulfur diesel becomes available, are high-usage commercial diesel vehicles that are on the roads of urban centers in Pakistan and primarily used within the city. The health benefits of retrofitting per vehicle per year are estimated in this book to be in the range of about US$95–568 for a DOC and US$216–1,295 for a DPF, depending on the type of vehicle and annual usage (table 4).

A DOC costs US$1,000–2,000 and a DPF as much as US$6,000–10,000. Therefore, the expected number of years that the vehicle will continue to be in use and years that the devices will be effective is an important consideration.

The midpoint estimate of the benefit-cost ratio of retrofitting in-use diesel vehicles with a DOC, once 500-ppm diesel is available, is about 1–1.3 for large buses and trucks used within the city, but the benefit is lower than the cost for minibuses and light-duty vans (figure 7).[16] Estimated health benefits of a DPF are currently lower than its cost for all classes of diesel vehicles, but should be reassessed once 50-ppm sulfur diesel is available in the future.

Given the relatively high cost of DOCs per unit of PM emission reduction, alternative options can be considered for in-use diesel-fueled minibuses and

Table 4 Estimated Health Benefits of Retrofitting Diesel Vehicles with PM Control Technology
US$/vehicle/year

Vehicle type	Diesel oxidation catalyst Vehicle usage		Diesel particulate filter Vehicle usage	
	35,000 km/year	70,000 km/year	35,000 km/year	70,000 km/year
Heavy-duty trucks	284	568	647	1,295
Large buses	208	417	475	949
Minibuses	133	265	302	604
Light-duty vans	95	189	216	432

Source: Sánchez-Triana and others 2014.

Figure 7 Benefit-Cost Ratios of Retrofitting In-Use Diesel Vehicles with DOC

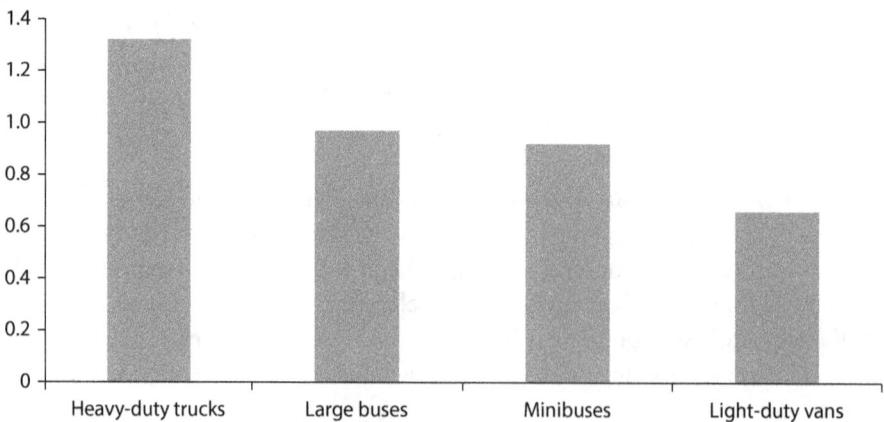

Source: Sánchez-Triana and others 2014.
Note: Seven years of useful life of diesel oxidation catalysts (DOC); 50,000 km/year vehicle usage.

Table 5 Estimated Health Benefits of Conversion to CNG
US$/vehicle/year

	Vehicle usage	
Vehicle type	35,000 km/year	70,000 km/year
Minibuses	644	1,288
Light-duty vans	455	909

Source: Sánchez-Triana and others 2014.
Note: CNG = compressed natural gas.

light-duty vans. One such option is conversion to CNG, which almost entirely eliminates PM emissions. This book estimates the health benefits of CNG conversion to be in the range of about US$455–1,288 per vehicle per year, depending on the type of vehicle and annual usage (table 5).

Conversion of such vehicles to CNG in Pakistan is reported to cost in the range of PRs 150–200 thousand per vehicle, or US$1,900–2,550 at exchange rates of 2009. Applying a cost of PRs 200,000, the midpoint estimate of the benefit-cost ratios on conversion to CNG are 1.7 for minibuses and 1.2 for light-duty vans.[17] The benefit-cost ratios for vans are somewhat lower than for minibuses due to a difference in estimated PM emissions per kilometer of vehicle use.

Health benefits can be twice the cost of converting to CNG... Two-stroke rickshaws are also a large source of PM emissions and urban noise. Conversion to a four-stroke engine using CNG is an option and is reported to cost around PRs 40–60 thousand in Pakistan. At a cost of PRs 60,000, the midpoint estimate of health benefits is two times the conversion cost. Two-stroke motorcycles also tend to have high PM emissions per ton of gasoline consumption and contribute greatly to urban noise. Many countries are therefore limiting or banning the use of two-stroke motorcycles. Four-stroke motorcycles emit substantially less PM, are more fuel efficient, and cause substantially less noise. The additional cost of a four-stoke motorcycle engine is approximately US$50 but varies with engine size. At this cost, and for a two-stroke motorcycle emitting 0.25 grams of PM per kilometer, the health benefits of switching to a four-stroke motorcycle are around US$2.9 per dollar of additional engine cost. When fuel savings are taken into account, the benefit-cost ratio is as high as 4.2.

Restrict CNG in autos... In view of CNG supply constraints and more energy-efficient alternative uses of CNG (for example, thermal power generation), it would be advisable to restrict CNG in automobiles (which make ubiquitous use of inefficient CNG conversion kits) and reserve its use for commercial and public service vehicles (buses, vans, utility trucks, and rickshaws). Emissions from spark-ignition automobiles and other vehicles could be better controlled by requiring Euro 2 and higher emission standards on both locally assembled and imported automobiles (including used cars), now that unleaded gasoline is locally refined and is available in the country.

Concerns about CNG... However, concerns about safety and economic reliability might threaten the sustainability of CNG substitution. A number of factors can affect the economic attractiveness of transport fueled by natural gas. These factors include the availability of natural gas, the high cost of converting vehicles to CNG, continuing public concerns about safety (for example, conversions of poor quality have led to some safety hazards, such as bus and taxi fires in some cities around the world), and limited incentives to build CNG fueling stations (Gwilliam, Kojima, and Johnson 2004, 36). Switching to natural gas would make economic sense only if the retail price of CNG were about 55–65% of the cost of the fuel being replaced (Kojima and others 2000, 4). The GoP might consider developing long-term agreements with neighboring countries to ensure a steady supply of natural gas before promoting further conversion efforts.

Mass transportation and traffic management... In the medium term, the GoP might consider other alternatives to reduce air pollution, particularly from mobile sources, such as supporting the development of mass transportation and improved traffic management in Pakistan's main cities. Experiences from countries such as Brazil, Colombia, and Mexico demonstrate the benefits of relatively new public transport systems, such as Bus Rapid Transit, which can use bus-based technologies to transport increasingly larger volumes of customers at moderately high speeds even in very congested urban areas. While still substantial, the investments needed to develop and operate these systems are significantly lower than those of traditional mass transport systems, such as underground metros. In addition, these systems have been able to demonstrate their contributions to reduce congestion and pollution, and some of them have even received international funding for their role in reducing GHG emissions from mobile sources. Additional policies are worth assessing in the medium term. They include traffic control, restricted circulation of private cars during high pollution episodes, urban planning and land use, establishment of high occupancy vehicle lanes, measures to improve traffic flow such as 'green wave' coordination of traffic signals, and improvement of infrastructure (for example, paving of roads and regular sweeping would reduce PM emissions).[18]

Non-motorized vehicles... The use of non-motorized vehicles, such as bicycles, should also be promoted as a means to reduce pollution from mobile sources. However, in doing so, the GoP might consider the development of dedicated bike paths, pedestrian zones, and other measures to reduce the potential conflict between motorized and non-motorized vehicles.

Policy Options for Abating Urban Air Pollution from Stationary Sources

Power plants, burning of waste... Stationary sources of air pollutant emissions in Pakistan include thermoelectric power plants; industrial units; and the burning of agricultural residues, sugarcane fields, and municipal waste. According to the 2011 Census of Manufacturing Industries, the number of industrial facilities in Pakistan amounts to 6,417 industrial units. Textiles, food products and beverages, tanneries, and chemical industries account for 60% of the total. Most of the industrial facilities are located in clusters in Punjab and Sindh.

Self-reported data... Most data on the emission of pollutants by stationary sources, such as factories, come from a database for self-reported pollution information. The federal government issued the NEQS for the Self-Monitoring and Reporting Tool (SMART) in 2001, under which all industrial units are to submit reports required for the industry category in which their unit is placed. By mid 2014, SMART had proven unable to provide information on emission of pollutants by stationary sources in Pakistan. The level of reporting on air pollutant emissions is low, as only 99 out of 6,417 industrial facilities have registered their emissions under the SMART program.

Industrial activities... Industrial activities, particularly those using fossil fuels and those manufacturing cement, are a significant source of air pollution. Despite the paucity of data, sporadic monitoring of air pollutants in Pakistan's industrial areas suggests that international standards for PM_{10} and NO_X are exceeded frequently (table 6) (Hashmi and Khani 2003; Hashmi, Shaikh, and Usmani 2005; Ministry of Finance 2010; Rajput and others 2005; Smith and others 1996; Wasim and others 2003). Key binding constraints to controlling air pollution in Pakistan include gaps in identification of stationary sources, an ineffective voluntary program for monitoring air emissions from stationary sources, a weak regulatory framework, and lack of enforcement.

Furnace oil... High sulfur content of furnace oil should be reduced, as it contributes significantly to PM and SO_2 emissions. Sulfur concentrations in furnace oil used in Pakistan electricity generation and industrial applications range from 2,000 to 9,000 ppm.

Table 6 Fuel and Air Pollution Emissions

Fuel type	$PM_{2.5}$	SO_x	NO_x	VOCs
Propane		M	L	M
Natural gas	L	M	H	M
Kerosene	M	H	M	M
Diesel (low sulfur)	M	M	H	M
Diesel (high sulfur)	H	H	H	M
Fuel oil	H	H	H	M
Crude oil	H	H	H	M
Coal (low sulfur)	H	M	M	L
Coal (high sulfur)	H	H	M	L
Wood and biomass	H	H	H	H

Source: Adapted from Sánchez-Triana 1993.
Note: L = low concentration of atmospheric emissions, M = medium concentration of atmospheric emissions, H = high concentration of atmospheric emissions, $PM_{2.5}$ = particulate matter of less than 2.5 microns, SO_x = sulfur oxides, NO_x = nitrogen oxides, VOC = volatile organic compounds. Blanks indicate no atmospheric emission.

Diffuse sources of air pollution... Diffuse sources of air pollution in Pakistan include the burning of municipal waste and of bagasse from sugarcane fields in rural areas, and the formation of dust clouds due to dry meteorological conditions. Due to their heterogeneity, nonpoint source pollution is more difficult to monitor and regulatory measures are more difficult to enforce.

Tackling diffuse sources of air pollution... Several measures and proper institutional coordination are still needed to tackle diffuse sources of air pollution in Pakistan. Solid waste collection by government-owned and government-operated services in Pakistan's cities currently averages only 50% of waste quantities generated; however, for cities to be relatively clean, at least 75% of these quantities should be collected. None of the cities in Pakistan has a proper solid waste management system extending from collection of solid waste up to its proper disposal. In terms of emissions stemming from the burning of fields for planting, green cut agricultural practices can make important contributions, as evidenced from experiences in Brazil.

Recommended measures... The recommended measures adopted by the GoP in the PCAP to address dispersed area sources are (a) block tree plantation; (b) afforestation in deserts; (c) sand dune stabilization; (d) paving of shoulders along main roads; and (e) proper disposal of solid waste. Implementing these measures will require a coordinated approach between the CCD, provincial EPAs, and other departments and local governments (Ghauri 2010).

Interventions targeting industrial sources... Interventions to control industrial air pollution in Pakistan include the replacement of fuel oil and coal by gas, introduction of low-sulfur diesel and furnace oil, energy efficiency, and end-of-pipe pollution control technology. Polluting factories could also move out from areas where they violate land-use regulations. Various environmental policy instruments are available to Pakistan for promoting interventions to control industrial air pollution. These include (a) economic and market-based instruments, (b) direct regulation by government or "command-and-control" measures, (c) public disclosure, and (d) legal actions.

Market-based instruments... Market-based instruments—such as taxes, pollution charges, tradable permits, or pricing policies—can be effective in moving existing markets and policies towards improved environmental outcomes. Economic instruments are based on the polluters pay principle where the polluting party pays for the damage done to the natural environment. A strong enforcement system is key to enhancing the effectiveness of economic instruments (World Bank 2008b).

Lack of understanding of the polluters pay principle... In Pakistan, the lack of understanding between the GoP and industries regarding the application of the PPP hampers the potential of market-based instruments for moving existing markets and policies towards improved environmental outcomes. In addition, a weak environmental framework, especially the failure to stipulate whether industries are responsible for bearing the cost of pollution reduction, hinders the successful implementation of cleaner production systems in heavily polluting industries.

Price distortions... Removal of price distortions in Pakistan could be designed to promote changes in the fuel mix towards cleaner fuels such as natural gas, while also contributing to moderating energy demand and reducing GHG emissions. Fuel prices are distorted because of rationing of fuel supply, but also because they fail to account for the adverse health effects of fossil fuel with the result that highly polluting fossil fuels are used excessively. Numerous examples demonstrate the perverse incentives linked to price distortions for energy, for example, disincentives to switch from diesel or furnace oil to natural gas. Another illustration involves the disincentive to energy conservation provided by heavily subsidized prices for gas to households and fertilizer plants. If price distortions to gasoline and diesel were removed, and the differences in tax rates for different fuels were eliminated, domestic prices for those fuels could be brought in line without major distortions. Natural gas would then become more competitive and there would be incentives for Pakistani firms to engage in international trade of natural gas, including imports from the Islamic Republic of Iran or Turkmenistan.

Shifting between fuels... In addition, pollution charges could be designed to promote a shift from using highly polluting fuels such as charcoal, biomass, or high-sulfur-content furnace oil and diesel to cleaner fuels such as natural gas. Differentiating prices across fuel types to minimize negative impacts on economic activity requires careful planning. An essential input for fuel taxes or pollution charges design is the targeting of fuels according to pollutants. Fuel types vary in their potential to produce atmospheric emissions that pollute the environment (table 6).

Command and control measures... Command and control measures can be effectively promoted and aligned with country context and conditions. This approach for preventing environmental problems generally relies on emissions standards, ambient standards, and technological standards, in conjunction with enforcement programs. Parameters for which limits have been established and regulated worldwide include pollutants such as $PM_{2.5}$ or PM_{10} (Faiz, Weaver, and Walsh 1996; Gurjar, Molina, and Ojha 2010). For example, command and control instruments are used in several countries for setting ambient air primary standards for $PM_{2.5}$ at 14.0 $\mu g/m^3$ (annual average) and 35 $\mu g/m^3$ (24-hour average).

Examples of common command and control instruments that use technological standards include phased reduction of the sulfur content in diesel to 500 ppm in the short term and to 15 ppm in the long term. Other technological standards include requiring relocation of industries from areas where ambient air pollution concentrations exceed the legal limits to places with no thermal inversions, good atmospheric mixing conditions, and outside areas of high-population density. The relocation of industries has also been used in many neighboring countries, such as China and India, as an opportunity to upgrade the plants' technology and improve environmental management in new plants (Li and Bin 2010).

One starting point... Pakistan could start by setting and monitoring standards for fine ($PM_{2.5}$) and ultrafine (PM_1) PM, given their impact on human health. A national apex organization on AQM could require action plans setting minimum air quality standards and proposing interventions for major cities.

An urgent need... Stationary sources, including point and nonpoint sources, are significant contributors to air pollution in Pakistan's urban areas. Currently, data on emissions from these sources are not systematically collected. At the operational level, there is the urgent need for collection of primary air quality information, and follow-up investigation in both the short and the medium term. This action includes improving the quality of data collection locally on the site, including training personnel. Data for SO_2, NO_2, O_3, and CO as well as PM_1, $PM_{2.5}$, and PM_{10} should continue to be recorded and coverage should be improved, in order to better understand the sources causing poor air quality. There is also an urgent need to carry out in-depth studies with instruments that are more sophisticated in order to better address the chemical and physical properties of the PM, especially in the fine mode.

Inventory of sources... Developing an inventory of stationary air pollution sources is an indispensable pillar of AQM. Official data indicate that only around 1% of the country's industrial establishments report their emissions under the SMART program. As a result, there is little to no information that can help the Pak-EPA or the provincial environmental agencies to map the location of different sources of air pollution, assess the types and quantities of pollutants that are being discharged into the environment, identify non-compliant units, and develop the necessary corrective actions. The inventory should initially focus on large polluters but gradually evolve to include small and medium enterprises, which may cumulatively emit large quantities of air pollutants.

Mandatory versus voluntary programs... Voluntary programs, including SMART, should be phased out and replaced by mandatory instruments that obligate polluters to comply with clear and enforceable environmental standards. Evidence from around the world, including both developed and developing countries, indicates that voluntary programs are ineffective for controlling pollution (Esty and Porter 2001; Morgenstern and Pizer 2007). In Pakistan, the extremely low number of units that participate in SMART illustrates the tool's limitations and its ineffectiveness to generate the information and actions needed to improve air quality.

Boosting competitiveness... Cleaner production can help boost Pakistan's industrial competitiveness.[19] Cleaner production programs aim at increasing firms' competitiveness and efficiency by helping them save energy, conserve water, control pollution, ensure safety of machines and equipment, improve health and safety of workers, improve environmental conditions, and enhance the image of the firm at the local and international levels. A World Bank consultant, Ralph Luken (2009), found that reducing emissions from industrial sources in Pakistan is financially viable and affordable.[20] Some cleaner technology interventions involve tuning and improving the maintenance of boilers and generators. Others involve technologies to remove PM from air emissions stacks. Many large firms in Pakistan in the textile and leather sectors are already undertaking energy efficiency measures (Khan 2011; Laiq 2010). Cleaner production systems and recycling of wastes can also reduce emissions from burning of municipal and agricultural wastes.

Mutually reinforcing phenomena... Climate change and air pollution are mutually reinforcing phenomena. As a result, there are important opportunities to build synergies in responses to both of these problems. Growing voluntary and compliance markets may provide financial resources for mitigating GHG emissions by undertaking actions that also improve urban air quality. Taking advantage of these opportunities will require aligning Pakistan's responses to both of these challenges and developing the institutional capacity to access such markets.

Capacity Building for Air Quality Management

Greater severity... Current urbanization, industrialization, and motorization trends in Pakistan suggest that, in the absence of targeted interventions, urban air pollution could become even more severe. In response to this challenge, the PCAP can undertake a number of interventions to strengthen institutional capacity for AQM. Priority interventions include enforcing ambient quality standards; enhancing air quality monitoring networks; regulating fuel quality; updating and enforcing vehicle emissions standards; and enforcing environmental standards for stationary source emissions. All have been successfully piloted both within and outside the region. Some would pose greater political difficulties and some would be more expensive, but all are cost-effective paths towards improved urban air quality.

Capacity building... Capacity-building efforts need to accompany all interventions for air pollution control. The stringency and economic efficiency of Pakistan's environmental regulatory framework and its effective enforcement are essential to improving air quality. At present, the penalties for noncompliance appear to be extremely modest. A more serious penalty system with strong governmental oversight could be adopted. Additionally, the organizational restructuring proposed by Pak-EPA requires specialists who can carry out a range of actions, including monitoring, enforcement, and planning. Recruiting the staff with the necessary expertise and background will be paramount to ensure that technical cells can fulfill their responsibilities.

Devolution of responsibilities... The 2010 modifications in Pakistan's constitution devolved major responsibilities for environmental management to subnational governments; this will have significant implications for air pollution control. Improving air quality in Pakistan will require the establishment of a central apex organization responsible for intersectoral and intergovernmental (national and provincial) coordination. Experiences from around the world, including both developed and developing countries, indicate that the existence of an apex organization is necessary to ensure coordination among different economic sectors, as well as across different levels of government. While decentralization of environmental management responsibilities offers a number of benefits, such as the capacity to respond more effectively to local priorities, there are also significant tradeoffs and risks. For example, unequal standards and institutional capacity could lead to more severe environmental degradation in different parts of the country.

Maintaining some responsibilities centrally... Even countries that use decentralized environmental management approaches maintain responsibilities impossible to delegate to regional entities. Some of the responsibilities are maintained centrally for reasons of efficiency, but others are kept as central responsibilities because failure to do so would be potentially harmful to the environment and the population. Specifically, the responsibilities that tend to be maintained by the central government, regardless of the level of decentralization, deal with (a) enacting environmental standards and policies; (b) national and international transboundary issues, including international agreements; (c) coordination between local governments; and (d) research into environmental issues, such as climate change adaptation and mitigation (Environmental Law Institute 2010).

Courts and judicial bodies, citizen enforcement, public interest advocacy... Strengthening enforcement of air quality regulations and the capacity of the legal courts and judicial bodies should not be postponed any longer. The 18th Constitutional Amendment's devolution of environmental issues to the provincial governments opened the opportunity to strengthen environmental tribunals (ETs) and create a more solid legal framework that would result in adequate sanctions being imposed on violators of environmental laws. The judicial system should be strengthened to act as an additional recourse for resolving environmental conflicts when enforcement fails. At least one ET should be created in each province and in Islamabad. This would imply creating three new ETs in addition to those established in Karachi and Lahore. In addition, taking advantage of the powers devolved to the provinces by the 18th Amendment to legislate on environmental matters, existing and future laws should explicitly provide for citizen enforcement. Finally, public interest advocacy is a powerful force for improvements in environmental management and should be supported through environmental law associations and the establishment of environmental law clinics at universities.

Conclusions and Recommendations

Harm from urban air pollution... The harm caused by air pollution in Pakistan's urban areas is one of the highest in the South Asia Region and exceeds several high-profile causes of mortality and morbidity in Pakistan. Urban air pollution silently kills more than 20,000 people each year and is responsible for more than 80,000 hospital admissions per year, including cases of chronic bronchitis, and several million cases of lower respiratory cases among children under five. Despite the strong evidence indicating an urgent need to improve urban air quality, the issue has received little attention and is yet to be included as a priority in Pakistan's policy agenda. In addition, there is little information on the main sources of lead and toxic pollutants, which should be identified in the short term, to be used as a basis for the development of targeted interventions.

Cannot manage without measuring... The old adage that it is impossible to manage what is not measured is particularly applicable to air quality. The monitoring equipment and assistance from JICA established a basic capability for monitoring air quality. However, ideally, the current monitoring program would be expanded and ongoing training and budget would be provided to the people who are maintaining and utilizing the equipment. The modest amount of actual monitored data presented in this book reflects the limited data available. Specialized equipment, regular maintenance, supplies of consumables, standardized protocols for reading and interpreting the data, and training of personnel are all essential elements of viable control policies. Furthermore, a centralized depository might be established to review and analyze data from across the country. Overall, a strong institutional structure, combined with solid base funding and continued oversight, is essential for achieving reliable and consistent measurements over time. In addition to installing and operating a reliable air quality monitoring network, inventories of mobile and stationary sources and modeling efforts are needed to better understand the contribution of different sources of pollution, including natural sources. These efforts should include efforts for source apportionment and speciation of $PM_{2.5}$ to better understand their source composition.

Consolidation... Consolidating the regulatory framework for air pollution control is indispensable to achieve improved urban air quality. The NEQS for ambient air quality published in late 2010 provide an important first step in establishing maximum limits that are consistent with research findings that document the health impacts of exposure to different air pollutants. However, the new NEQS still allow for concentrations of some pollutants at levels higher than those recommended by the WHO. A different, but also significant challenge will be enforcing the new standards, including compliance with the various deadlines.

Command and control instruments... Command and control instruments can be effective to control air pollution. For air pollution control, the most common command and control measures include ambient standards, emission standards, and technology-based and performance-based standards. In the short term, in Pakistan, the primary emphasis should be on reducing levels of pollutants linked to high morbidity and mortality: $PM_{2.5}$ (and precursors like sulfur oxides [SO_x] and nitrogen oxides [NO_x]) from mobile sources. A second level of priority can be given to $PM_{2.5}$, SO_x, and emissions of Pb (and other toxic metals) from stationary sources. A third level of priority could be given to O_3 and its precursors NO_x and VOCs (particularly air toxics like benzene). A fourth level of priority could be given to other traditional air pollutants such as CO and GHGs.

Pollution charges... Pollution charges could be used as an efficient mechanism to reduce pollution emissions from different sources. The provisions in PEPA on the establishment of pollution charges and the detailed formulas that were developed in 2001 for reporting and paying pollution charges can support their establishment. However, as with any other instrument established by law, pollution charges will not be effective if they are not strictly enforced.

Economically efficient interventions... Economically efficient interventions for reducing air pollution from mobile sources include, among others, the following:

- Continue moving to 500-ppm sulfur diesel in the short term and to 50 ppm in the medium term.
- Advance efforts to convert diesel-fueled minibuses and city delivery vans to CNG and install DOCs on existing large buses and trucks used in the city.
- Sustain the introduction of new CNG full-size buses, as DPFs cannot be used with 500-ppm sulfur in diesel.
- Convert existing two-stroke rickshaws to four-stroke CNG engines.
- Ban new two-stroke motorcycles and rickshaws and explore options to control PM emissions from in-use two-stroke motorcycles.
- Introduce low-sulfur fuel oil (1% sulfur) to major users located in urban centers.
- Control emissions from large point sources.
- Restrict use of CNG in spark-ignition automobiles and introduce Euro 2 and higher standards on locally assembled and imported private automobiles and light-duty utility vehicles.

Coordinated interventions in five areas... Policy makers in Pakistan face an array of obstacles, including limited financial, human, and technical resources and can only pursue a small number of strategic interventions on AQM at the same time. Addressing Pakistan's severe urban air pollution problem will require undertaking a series of coordinated interventions to strengthen air quality monitoring, build the institutional capacity of responsible agencies, bolster the legal and regulatory framework for AQM, carry out targeted policy reforms and investments, and fill existing knowledge gaps.

Building constituencies... Building environmental constituencies through dissemination of information can be an essential AQM building block in Pakistan. Mechanisms to disseminate information that is easily understood can empower communities to function as informal regulators and promote accountability on those being regulated. Examples are the pioneering public disclosure schemes in Colombia and Indonesia, where governments required industries to report their pollutant emissions and rate themselves on compliance with national standards. These ratings were then released publicly; those receiving failing grades were shamed, while those demonstrating sound environmental performance were publicly praised by government officials, NGOs, and the press. Additional opportunities include a public information program to support clean air through the public provision of air quality information. For instance, the publication of an Air Quality Index in major cities would build support for air quality improvement initiatives and enable the issuance of health alarms when necessary (Ahmed and Sánchez-Triana 2008; Blair 2008; Sánchez-Triana and Ortolano 2005; World Bank 2005). Table 7 summarizes key recommendations emerging from the present book.

Table 7 Recommended Short-Term Actions to Strengthen AQM in Pakistan

AQM management building block	Recommended action
Strengthening Air Quality Monitoring	Establish a reliable air quality monitoring network focusing on pollutants such as $PM_{2.5}$, SO_2, NO_3, and Pb, building on the network developed with JICA support, and providing ongoing training and budget for maintaining and utilizing the equipment.
	Develop a detailed mobile source emissions inventory.
	Establish an inventory of stationary sources, focusing on major industrial polluters and evolving to include small and medium enterprises.
	Carry out modeling efforts to assess the present and future contributions of mobile, stationary, nonpoint, and natural sources of key pollutants.
Building Institutional Capacity for AQM	Establish a central apex organization with a clear mandate for AQM and with responsibilities for intersectoral and intergovernmental (national and provincial) coordination.
	Strengthen provincial EPAs to carry out monitoring, enforcement, and planning activities related to AQM.
	Strengthen enforcement of air quality regulations and the capacity of the legal courts and judiciary to enforce environmental laws.

table continues next page

Table 7 Recommended Short-Term Actions to Strengthen AQM in Pakistan *(continued)*

AQM management building block	Recommended action
Bolstering the Legal and Regulatory Framework for AQM	Adopt pollution charges targeting fuels according to their pollution contributions, based on the PPP.
	Adopt a more serious penalty system for noncompliance with air quality laws and regulations.
	Phase out voluntary regulatory schemes and introduce enforceable standards for key stationary sources of air pollutants.
	Adopt revised NEQS to ensure that the permissible ambient concentrations of all pollutants are consistent with the levels recommended by the WHO.
Policy Reforms and Investments for Improved Air Quality	Reduce fuel price distortions and apply the PPP to promote efficient fuel use and switching towards gas and other cleaner fuels.
	Improve fuel quality by importing cleaner diesel and furnace oil, using low-sulfur crude oil, and investing in refinery capacity for desulfurization.
	Increase solid waste collection and proper disposal, particularly in large urban areas.
	Implement green cut agricultural practices to reduce emissions from nonpoint sources.
	Promote cleaner production in the industrial sector with the dual aim of improving environmental conditions and strengthening firm competitiveness.
	Foster the creation of a strong air quality constituency by providing training and disseminating specific materials among policy makers, legislators, NGOs, journalists, and other stakeholders.
	Adopt a public disclosure scheme requiring industries to report their pollutant emissions and rate themselves on compliance with national standards.
	Publish an Air Quality Index in major cities and issue health alarms when necessary.
Filling Knowledge Gaps for AQM	Carry out an in-depth study to identify source of pollution for Pb and other toxic substances to serve as basis for the development of targeted interventions.
	Conduct in-depth analysis of a potential scrappage program for old vehicles.
	Promote the establishment of research programs in universities with a focus on the different areas of AQM, including, but not limited to, law, economics, meteorology, and chemistry.

Note: AQM = air quality management; EPAs = Environmental Protection Agencies; NEQS = National Environmental Quality Standards; NGO = nongovernmental organization; $PM_{2.5}$ = particulate matter of less than 2.5 microns; PPP = polluter pays principle; WHO = World Health Organization.

Notes

1. Estimates from the Global Model of Ambient Particulates indicate that as of 2002, average PM_{10} in Pakistan was 165 µg/m^3 (World Bank 2006). These estimates were based on an econometrically estimated model for predicting PM levels.

2. Gurjar and others (2008) suggested a multi-pollutant index, which takes into account the combined levels of the three WHO criteria pollutants; they found that Karachi, one of Pakistan's mega-cities, held the fourth position worldwide on the multi-pollutant index-based ranking.

3. Recent advances in remote sensing have provided new avenues for measuring, monitoring, and understanding processes that lead to atmospheric pollution. Mansha and Ghauri (2011) assessed the seasonal and spatial variation of aerosol concentration over Karachi, Pakistan, using satellite-derived aerosol optical depth (AOD) data, as well as other ground-based data, including $PM_{2.5}$. It was demonstrated that satellite information can be used with a reasonable degree of accuracy, particularly when limited ground-based measurements are available. Recently, Alam and others (2010)

analyzed the spatio-temporal variation of AOD and the aerosol influence on cloud parameters using moderate resolution imaging spectroradiometer data; they found higher AOD values in summer and lower values in winter for various cities in Pakistan. Alam, Trautmann, and Blaschke (2011) studied the aerosol optical properties and radiative forcing over mega-city Karachi. The high July AOD values are predominantly caused by dust activities in the southern part of Pakistan, whereas high AOD values for November are due to high black carbon concentrations and to some dust plumes near Karachi (Dutkiewicz and others 2009). Alam and others (2012) again reported high AOD values over Lahore and Karachi during December, largely due to absorbing anthropogenic activities, while high AOD values in June were attributed to mineral dust aerosols. Alam, Qureshi, and Blaschke (2011) used three different satellite-borne sensors to investigate the spatial and temporal variations of aerosols over several cities in Pakistan. Back trajectory analyses indicated that while winter air masses reaching Pakistan had traveled long distances, summer air masses had traveled only short distances. While monsoonal rainfall tends to reduce PM concentrations by washing aerosols out of the atmosphere, this effect was found to be mainly restricted to Pakistan's eastern and southeastern parts.

4. Road safety data can be found at http://www.who.int/violence_injury_prevention /road_safety_status/ country_profiles/pakistan.pdf.

5. The figures refer to motor vehicles on the road, which, according to the Ministry of Finance (2012), are significantly more than the number of registered vehicles.

6. The results reported by Ghauri (2010) and Khan (2011) are part of a World Bank-financed study that present the data obtained by the consultant through a survey and interviews among different industrial units and other stakeholders in Pakistan.

7. Ambient air quality in Lahore has documented 24-hour maximum wintertime $PM_{2.5}$ concentrations of 200 µg/m³ (Biswas, Ghauri, and Husain 2008; Zhang, Quraishi, and Schauer 2008). Principal component analysis (PCA) showed that the greatest contributors to PM_{10} were industrial sources, followed by secondary aerosols and mobile sources.

8. The data on emissions sources and inventories in Pakistan need to be improved to refine the design and targeting of control strategies for both mobile and stationary sources.

9. Mortality is estimated based on Ostro (2004) and on Pope and others (2002, 2009).

10. Road safety data can be found at http://www.who.int/violence_injury_prevention /road_safety_status/ country_profiles/pakistan.pdf.

11. NEQS for Motor Vehicle Exhaust and Noise. These NEQS (also known in Pakistan as Euro 2 Standards) replace Annex III of the Notification S.R.O. 742 (I) 93, dated August 24, 1993. The S.R.O. 72 (KE)/2009 introduces a set of emission standards for all new and in-use vehicles. Those categories are subdivided into diesel (light and heavy) and gasoline (petrol)-powered vehicles (passenger cars, light commercial vehicles, rickshaws, and motorcycles). For example, for heavy diesel engines and large-goods vehicles (both locally manufactured and imported), the standard for PM is 0.15 g/kWh, to be enforced after July 1, 2012 (the standard does not define if this value refers to PM_{10}, or $PM_{2.5}$). No reference is made to standards for SO_2. The schedule of implementation includes three deadlines: immediate; July 1, 2009; and July 1, 2012. The smoke, CO, and noise standards for in-use vehicles were to be effective immediately after the regulation was adopted. The S.R.O. 1062 (I)/2010 established NEQS for ambient air quality. Its schedule of implementation includes two deadlines: July 1, 2010, and January 1, 2013.

12. The existing air quality monitoring network collects data on total suspended particles (TSP), PM_{10}, NO_2, SO_2, and CO.

13. The emission standards for new vehicles are generally different from in-use vehicles, since the needs and test procedures are different. Euro 2 standards require 3-way catalysts. For heavy-duty diesel vehicles, PM and NO_x standards are specified together, since there is a trade-off between these two in terms of engine optimization.

14. Due to severe shortage of natural gas in the country, the government has introduced closure of CNG fuel stations in Punjab and Khyber Pakhtunkhwa for three months and up to 48 hours per week in the Sindh province, with no supplies to industries including powerhouses. Many CNG users have gone back to petrol for fueling, and automobile manufacturers who once introduced company-fitted CNG kits in their vehicles have withdrawn the facility. Because of these constraints, converting diesel-fueled minibuses and vans to compressed natural gas (CNG) should only be considered a viable policy if supplies from international or domestic producers are secured.

15. Nearly 90% of premature mortality from PM pollution in Karachi is among adults. These individuals may be losing around 10 years of life due to PM pollution. The book estimates the social cost of premature mortality using two values (that is, the human capital value and the value of statistical life) which, for adults, give vastly different cost estimates. The value of statistical life is applied in the assessment of benefits and costs of PM control interventions, because this value is more likely than the human capital value to be closer to the actual value that individuals place on a reduction in the risk of death.

16. A cost of a DOC of US$1,500 was applied for heavy-duty trucks and large buses, and a cost of US$1,000 for minibuses and light-duty vans. A discount rate of 10% was applied to annualize the cost of the DOC.

17. A discount rate of 10% was applied to annualize the cost of conversion to CNG.

18. By 2013, large cities in Pakistan like Karachi, Lahore, and Rawalpindi introduced 'Signal-free Corridors' on congested and heavy traffic-load city road sections, which helped to avoid long waiting times. Maintaining optimal speeds helps to reduce vehicular emissions.

19. According to the United Nations Environment Programme, cleaner production is "the continuous application of an integrated environmental strategy to processes, products and services to increase efficiency and reduce risks to humans and the environment" (Luken 2009, 9).

20. At selected paper mills, implementation of energy-efficient measures resulted in a 5% reduction of electricity and fuel consumption (Khan 2011). In paper manufacturing factories, cleaner production also helped eliminate chlorine emission from the process. Improvement in the bleaching process resulted in a 10% reduction in chemical consumption (Khan 2011).

Bibliography

Ahmed, K., and E. Sánchez-Triana. 2008. *Strategic Environmental Assessment for Policies: An Instrument for Good Governance*. Washington, DC: World Bank. https://openknowledge.worldbank.org/handle/10986/6461 License: Creative Commons Attribution CC BY 3.0.

Akbar, S., and K. Hamilton. 2010. "Assessing the Environmental Co-Benefits of Climate Change Actions." 2010 Environment Strategy, Analytical Background Papers, World Bank, Washington, DC.

Alam, K., T. Blaschke, P. Madl, A. Mukhtar, M. Hussain, T. Trautmann, and S. Rahman. 2011. "Aerosol Size Distribution and Mass Concentration Measurements in Various Cities of Pakistan." *Journal of Environmental Monitoring* 13: 1944–52.

Alam, K., M. J. Iqbal, T. Blaschke, S. Qureshi, and G. Khan. 2010. "Monitoring Spatio-Temporal Variations in Aerosols and Aerosol-Cloud Interactions over Pakistan Using MODIS Data." *Advances in Space Research* 46: 1162–76.

Alam, K., S. Qureshi, and T. Blaschke. 2011. "Monitoring Spatio-Temporal Aerosol Patterns Over Pakistan Based on MODIS, TOMS and MISR Satellite Data and a HYSPLIT Model." *Atmospheric Environment* 45: 4641–51.

Alam, K., T. Trautmann, and T. Blaschke. 2011. "Aerosol Optical Properties and Radiative Forcing Over Mega City Karachi." *Atmospheric Research* 101: 773–82.

Alam, K., T. Trautmann, T. Blaschke, and H. Majid. 2012. "Aerosol Optical and Radiative Properties during Summer and Winter Seasons over Lahore and Karachi." *Atmospheric Environment* 50 (April): 234–45. doi: 10.1016/j.atmosenv.2011.12.027.

Ali, M., and M. Athar. 2010. "Impact of Transport and Industrial Emissions on the Ambient Air Quality of Lahore City, Pakistan." *Environmental Monitoring and Assessment* 171 (1–4): 353–63.

Aziz, A., and I. U. Bajwa. 2004. "Energy and Pollution Control Opportunities for Lahore." *Urban Transport X* 16: 751–60.

———. 2007. "Minimizing Human Health Effects of Urban Air Pollution through Quantification and Control of Motor Vehicular Carbon Monoxide (CO) in Lahore." *Environmental Monitoring and Assessment* 135: 1–3.

———. 2008. "Erroneous Mass Transit System and Its Tended Relationship with Motor Vehicular Air Pollution (An Integrated Approach for Reduction of Urban Air Pollution in Lahore)." *Environmental Monitoring and Assessment* 137: 25–33.

Aziz, J. A. 2006. "Towards Establishing Air Quality Guidelines for Pakistan." *Eastern Mediterranean Health Journal* 12 (6): 886–93.

Bhutto, A., A. A. Bazmib, and G. Zahedi. 2011. "Greener Energy: Issues and Challenges for Pakistan, Biomass Energy Prospective." *Renewable and Sustainable Energy Reviews* 15 (6): 3207–19.

Biswas, K. F., B. M. Ghauri, and L. Husain. 2008. "Gaseous and Aerosol Pollutants during Fog and Clear Episodes in South Asian Urban Atmosphere." *Atmospheric Environment* 42 (33): 7775–85.

Blair, H. 2008. "Building and Reinforcing Social Accountability for Improved Environmental Governance." In *Strategic Environmental Assessment for Policies: An Instrument for Good Governance*, edited by K. Ahmed and E. Sánchez-Triana, 127–57. Washington, DC: World Bank. https://openknowledge.worldbank.org/handle/10986/6461 License: Creative Commons Attribution CC BY 3.0.

Colbeck, I., Z. A. Nasir, and Z. Z. Ali. 2010a. "The State of Air Quality in Pakistan: A Review." *Environmental Science and Pollution Research* 17: 49–63.

———. 2010b. "The State of Indoor Air Quality in Pakistan: A Review." *Environmental Science and Pollution Research* 17: 1187–96.

Colbeck, I., Z. A. Nasir, Z. Ali, and S. Ahmed. 2010. "Nitrogen Dioxide and Household Fuel Use in Pakistan." *Science of the Total Environment* 409: 357–63.

Dall'Osto, M. 2012. Unpublished consultant report for the World Bank, Washington, DC.

Dutkiewicz, V. A., S. Alvi, B. M. Ghauri, M. I. Choudhary, and L. Hussain. 2009. "Black Carbon Aerosols in Urban Air in South Asia." *Atmospheric Environment* 43: 1737–44.

Environmental Law Institute. 2010. *India 2030: Vision for an Environmentally Sustainable Future. Best Practices Analysis of Environmental Protection Authorities in Federal States.* Study commissioned by the World Bank, Washington, DC. http://www.eli.org/pdf /india2030.pdf.

Esty, D. C., and M. E. Porter. 2001. "Ranking National Environmental Regulation and Performance: A Leading Indicator of Future Competitiveness?" In *The Global Competitiveness Report 2001*, edited by M. E. Porter, J. Sachs, and A. M. Warner, 78–100. New York: Oxford University Press.

Faiz, A. 2011. Comments and edits on Draft Air Quality Management Report. E-mail correspondence with authors, June 19, 26, 29, and 30.

Faiz, A., C. S. Weaver, and M. P. Walsh. 1996. "Air Pollution from Motor Vehicles: Standards and Technologies for Controlling Emissions." World Bank, Washington, DC.

Ghauri, B. 2008. "Satellite Data Applications in Atmospheric Monitoring." SUPARCO, presented at the United Nations/Austria/European Space Agency Symposium, Graz, Austria, September 9–12.

———. 2010. *Institutional Analysis of Air Quality Management in Urban Pakistan.* Consulting report commissioned by the World Bank, Washington, DC.

Ghauri B., A. Lodhi, and M. Mansha. 2007. "Development of Baseline (Air Quality) Data in Pakistan." *Environmental Monitoring and Assessment* 127: 237–52.

Gurjar, B. R., T. M. Butler, M. G. Lawrence, and J. Lelieveld. 2008. "Evaluation of Emissions and Air Quality in Megacities." *Atmospheric Environment* 42:1593–606.

Gurjar, B. R., L. T. Molina, and C. S. P. Ojha. 2010. *Air Pollution. Health and Environmental Impacts.* Boca Raton, FL: CRC Press.

Gwilliam, K., M. Kojima, and T. Johnson. 2004. *Reducing Air Pollution from Urban Transport.* Washington, DC: World Bank.

Hameed, S., M. I. Mirza, B. M. Ghauri, Z. R. Siddiqui, R. Javed, A. R. Khan, O. V. Rattigan, S. Qureshi, and L. Husain. 2000. "On the Widespread Winter Fog in Northeastern Pakistan and India." *Geophysical Research Letters* 27 (13): 1891–94.

Harrison R. M., C. Giorio, D. C. Beddows, and M. Dall'Osto. 2010. "Size Distribution of Airborne Particles Controls Outcomes of Epidemiological Studies." *Science of the Total Environment* 409: 289–93.

Harrison R. M., D. J. T. Smith, C. A. Pio, and L. M. Castro. 1997. "Comparative Receptor Modeling Study of Airborne Particulate Pollutants in Birmingham (United Kingdom), Coimbra (Portugal) and Lahore (Pakistan)." *Atmospheric Environment* 31 (20): 3309–21.

Harrison, R. M., and J. Yin. 2000. "Particulate Matter in the Atmosphere: Which Particle Properties Are Important for Its Effects on Health?" *Science of the Total Environment* 249: 85–101.

Hashmi, D. R., and M. I. Khani. 2003. "Measurement of Traditional Air Pollutants in Industrial Areas of Karachi, Pakistan." *Journal of the Chemical Society of Pakistan* 25:103–09.

Hashmi, D. R., G. H. Shaikh, and T. H. Usmani. 2005. "Air Quality in the Atmosphere of Karachi City: An Overview." *Journal of the Chemical Society of Pakistan* 27: 6–13.

Husain, L., V. A. Dutkiewicz, A. J. Khan, and B. M. Ghauri. 2007. "Characterization of Carbonaceous Aerosols in Urban Air." *Atmospheric Environment* 41 (32): 6872–83.

Husain, L., B. K. Farhana, and B. M. Ghauri. 2007. "Emission Sources and Chemical Composition of the Atmosphere of a Mega-city in South Asia." *Eos Trans AGU* 88 (23), Joint Assembly Supplement, A31C–01.

Hussain, A., H. Mir, and M. Afzal. 2005. "Analysis of Dust Storms Frequency over Pakistan during 1961–2000." *Pakistan Journal of Meteorology* 2 (3): 49–68.

IANGV (International Association for Natural Gas Vehicles). 2003. "Natural Gas Vehicles and Climate Change. A Briefing Paper." http://www.ruscom.com/ngvbc/downloads/altfuels/Briefing_paper.pdf.

IEA (International Energy Agency). 2009. *World Energy Outlook 2008.* Paris: IEA.

Ilyas, S. Z. 2007. "A Review of Transport and Urban Air Pollution in Pakistan." *Journal of Applied Science and Environmental Management* 11 (2): 113–21.

Khan, A. U. 2011. *Industrial Environmental Management in Pakistan.* Consulting report commissioned by the World Bank, Washington, DC.

Kojima, M., R. W. Bacon, M. Fodor, and M. Lovei. 2000. *Cleaner Transport Fuels for Cleaner Air in Central Asia and the Caucasus.* Washington, DC: World Bank.

Laiq, A. 2010. *Evaluation of Cleaner Production Programs in Pakistan.* Consulting report commissioned by the World Bank, Washington, DC.

Li, L., and L. Bin. 2010. *Environmental Cost Analysis of the Relocation of Pollution-Intensive Industries Case Study: Transfer of Ceramics Industry from Foshan to Qingyang, Guangdong Province.* Research Report 2010-RR2. Singapore: Economy and Environment Program for Southeast Asia.

Lodhi, A., G. Badar, M. K. Rafiq, S. Rahman, and S. Shoaib. 2009. "Particulate Matter ($PM_{2.5}$) Concentration and Source Apportionment in Lahore." *Journal of the Brazilian Chemical Society* 20 (10): 1811–20.

Luken, R. A. 2009. *Industrial Environmental Management in Pakistan.* Consulting report commissioned by the World Bank, Washington, DC.

Mansha, M., and B. Ghauri. 2011. "Assessment of Fine Particulate Matter ($PM_{2.5}$) in Metropolitan Karachi Through Satellite and Ground-based Measurements." *Journal of Applied Remote Sensing* 5. doi: 10.1117/1.3625615.

Mansha, M., B. Ghauri, S. Rahman, and A. Amman. 2011. "Characterization and Source Apportionment of Ambient Air Particulate Matter ($PM_{2.5}$) in Karachi." *Science of the Total Environment.* doi: 10.1016/j.scitotenv.2011.10.056.

Ministry of Finance. 2010. *Pakistan Economic Survey 2009–10.* Islamabad: Ministry of Finance, Government of Pakistan. http://www.finance.gov.pk/survey_0910.html.

———. 2012. *Pakistan Economic Survey 2011–12.* Islamabad: Ministry of Finance, Government of Pakistan. http://www.finance.gov.pk/survey_1112.html.

Morgenstern, R., and B. Pizer. 2007. *Reality Check. The Nature of Performance of Voluntary Environmental Programs in the United States, Europe, and Japan.* Washington, DC: Resources for the Future.

Ostro, B. 2004. *Outdoor Air Pollution: Assessing the Environmental Burden of Disease at National and Local Levels.* Environmental Burden of Disease, Series 5. Geneva: World Health Organization.

Pak-EPA. 2007. Ambient Air and Water Quality Investigation in Quetta. http://www
.environment.gov.pk/pub-pdf/Ambient%20AW%20Quetta.pdf.

Pak-EPA/JICA. 2001. "Cities Investigation of Air and Water Quality (Lahore, Rawalpindi
& Islamabad)." JICA–Pak-EPA.

———. 2006. "Measurement of NO_2 Concentration in Ambient Air in Major Cities of
Pakistan Using Diffusion Samplers." JICA–Pak-EPA.

Parekh, P. P., B. Ghauri, and L. Husain. 1989. "Identification of Pollution Sources of
Anomalously Enriched Elements." *Atmospheric Environment* 23: 1435–42.

Pope, C. A., III, R. T. Burnett, D. Krewski, M. Jerrett, Y. Shi, E. E. Calle, and M. J. Thun.
2009. "Cardiovascular Mortality and Exposure to Airborne Fine Particulate Matter
and Cigarette Smoke: Shape of the Exposure-Response Relationship." *Circulation* 120:
941–48.

Pope, C. A., III, R. T. Burnett, M. J. Thun, E. Calle, D. Krewski, K. Ito, and G. Thurston.
2002. "Lung Cancer, Cardiopulmonary Mortality, and Long-term Exposure to Fine
Particulate Air Pollution." *JAMA* 287: 1132–41.

Quraishi, T. A., J. J. Schauer, and Y. X. Zhang. 2009. "Understanding Sources of Airborne
Water Soluble Metals in Lahore, Pakistan." *Kuwait Journal of Science and Engineering*
36 (1A): 43–62.

Raja, S., F. K. Biswaw, P. K. Hopke, and L. Husain. 2010. "Source Apportionment of
the Atmospheric Aerosol in Lahore, Pakistan." *Water, Soil, and Air Pollution* 208:
43–57.

Rajput, M. U., S. Ahmad, M. Ahmad, and W. Ahmad. 2005. "Determination of Elemental
Composition of Atmospheric Aerosol in the Urban Area of Islamabad, Pakistan."
Journal of Radioanalytical & Nuclear Chemistry 266 (2): 343–48.

Sami, M., A. Waseem, and S. Akbar. 2006. "Quantitative Estimation of Dust Fall and
Smoke Particles in Quetta Valley." *Journal of Zhejiang University SCIENCE B* 7 (7):
542–47. http://www.ncbi.nlm.nih.gov/pmc/articles/PMC1500881/?tool=pmcentrez.

Sánchez-Triana, E. 1993. *Mecanismos financieros e incentivos económicos para la protección
ambiental.* Bogotá: Friedrich Ebert Stiftung.

Sánchez-Triana, E., J. Afzal, D. Biller, and S. Malik. 2013. *Greening Growth in Pakistan
through Transport Sector Reforms: A Strategic Environmental, Poverty, and Social
Assessment.* Directions in Development. Washington, DC: World Bank. doi: 10.1596
/978-0-8213-9929-3.

Sánchez-Triana, E., S. Enriquez, B. Larsen, and E. Golub. 2014. *Environmental and Climate
Change Priorities for the Sindh Province.* Environment and Development Series.
Washington, DC: World Bank.

Sánchez-Triana, E., K. Ahmed, and Y. Awe. 2007. *Environmental Priorities and Poverty
Reduction: A Country Environmental Analysis for Colombia.* Directions in Development.
Environment and Sustainable Development. Washington, DC: World Bank.

Sánchez-Triana, E., and L. Ortolano. 2005. "Influence of Organizational Learning on
Water Pollution Control in Colombia's Cauca Valley." *International Journal of Water
Resources Development* 21 (3): 493–508.

Shafer, M. M., D. A. Perkins, D. S. Antkiewicz, E. A. Stone, T. A. Quraishi, and J. J. Schauer.
2010. "Reactive Oxygen Species Activity and Chemical Speciation of Size-fractionated
Atmospheric Particulate Matter from Lahore, Pakistan: An Important Role for
Transition Metals." *Journal of Environmental Monitoring* 12 (3): 704–15.

Shah, M., and H. N. Shaheen. 2007a. "Annual TSP and Tracemetal Distribution in the Urban Atmosphere of Islamabad in Comparison with Mega-cities of the World." *Human and Ecological Risk Assessment* 13: 884–99.

———. 2007b. "Statistical Analysis of Atmospheric Trace Metals and Particulate Fractions in Islamabad, Pakistan." *Journal of Hazardous Material* 147 (3): 759–67. doi: 10.1016/j.jhazmat.2007.01.075.

———. 2009. "Study of Particle Size and Trace Metal Distribution in Atmospheric Aerosols of Islamabad." *Journal of the Chemical Society of Pakistan* 31 (3): 427–33.

———. 2010. "Seasonal Behaviors in Elemental Composition of Atmospheric Aerosols Collected in Islamabad, Pakistan." *Atmospheric Research* 95 (2–3): 210–23. doi: 10.1016/j.atmosres.2009.10.001.

Shah, M., N. Shaheen, and R. Nazir. 2012. "Assessment of the Trace Elements Level in Urban Atmospheric Particulate Matter and Source Apportionment in Islamabad, Pakistan." *Atmospheric Pollution Research*. doi: 10.5094/APR.2012.003.

Sindh EPA. 2010. Daily $PM_{2.5}$ data collected from air quality monitoring network, Karachi. Unpublished.

Smith, D. J. T., R. M. Harrison, L. Luhana, C. A. Pio, L. M. Castro, M. N. Tariq, S. Hayat, and T. Quraishi. 1996. "Concentrations of Particulates Airborne Polycyclic Aromatic Hydrocarbons and Metals Collected in Lahore, Pakistan." *Atmospheric Environment* 30: 4031–40.

Stone, E., J. Schauer, T. A. Quraishi, and A. Mahmood. 2010. "Chemical Characterization and Source Apportionment of Fine and Coarse Particulate Matter in Lahore, Pakistan." *Atmospheric Environment* 44: 1062–70.

von Schneidemesser, E., E. A. Stone, T. Quraishi, M. M. Shafer, and J. J. Schauer. 2010. "Toxic Metals in the Atmosphere in Lahore, Pakistan." *Science of the Total Environment* 408 (7): 1640–48.

Wahid, A. 2006a. "Influence of Atmospheric Pollutants on Agriculture in Developing Countries: A Case Study with Three New Varieties in Pakistan." *Science of the Total Environment* 37 (1–3): 304–13.

———. 2006b. "Productivity Losses in Barley Attributable to Ambient Atmospheric Pollutants in Pakistan." *Atmospheric Environment* 40 (28): 5342–54.

Wasim, M., A. Rahman, S. Waheed, M. Daud, and S. Ahmad. 2003. "INAA for the Characterization of Airborne Particulate Matter from the Industrial Area of Islamabad City." *Journal of Radioanalytical and Nuclear Chemistry* 258 (2): 397–402.

WHO (World Health Organization). 2006. "Air Quality Guidelines, Global Update 2005." Regional Office for Europe, WHO.

———. 2009. *Global Status Report on Road Safety: Time for Action.* Geneva: WHO. http://whqlibdoc.who.int/publications/2009/9789241563840_eng.pdf.

World Bank. 2005. *Integrating Environmental Considerations in Policy Formulation: Lessons from Policy-Based SEA Experience.* Report 32783. Washington, DC: World Bank.

———. 2006. "Pakistan Strategic Country Environmental Assessment." Washington, DC: World Bank.

———. 2007. *Republic of Peru—Environmental Sustainability: A Key to Poverty Reduction in Peru.* Washington, DC: World Bank.

————. 2008a. *Environmental Health and Child Survival. Epidemiology, Economics, Experience.* Environment and Development. Washington, DC: World Bank. doi: 10.1596/978-0-8213-7236-4.

————. 2008b. *Senegal—Country Environmental Analysis.* Washington, DC: World Bank.

————. 2010. "World Development Indicators." http://data.worldbank.org.

————. 2011. *Policy Options for Air Quality Management in Pakistan.* Report delivered to the Government of Pakistan, Washington, DC: World Bank.

————. 2012. *Environmental Priorities and Climate Change in Sindh.* Report delivered to the Government of Sindh (Pakistan), Washington, DC: World Bank.

Zhang, Y. X., T. Quraishi, and J. J. Schauer. 2008. "Daily Variations in Sources of Carbonaceous Aerosol in Lahore, Pakistan during a High Pollution Spring Episode." *Aerosol and Air Quality Research* 8 (2): 130–46.

CHAPTER 1

Overview

Introduction

Pakistan's urban air pollution is among the most severe in the world and it engenders significant damages to human health and the economy. Pakistan is the most urbanized country in South Asia and it is undergoing rapid motorization and increasing energy use. Air pollution, particularly in large urban centers, damages the populations' health and quality of life and contributes to environmental degradation (Aziz 2006; Aziz and Bajwa 2004; Colbeck, Zaheer, and Zulfiqar 2010; World Bank 2006).[1] From 2007 to 2011, the reported levels of particulate matter (PM), sulfur dioxide (SO_2), and lead (Pb) were many times higher than the World Health Organization (WHO) air quality guidelines.[2,3]

Air pollution, inadequate water supply, sanitation, and hygiene are the top environmental priority problems in Pakistan. Damages are associated, mainly, with increased premature mortality and learning disability. Outdoor air pollution in Pakistan disproportionately affects the health and productivity of poor people, especially for the more than 35% of Pakistanis who live in urban areas. In 2005, more than 22,600 adult deaths were attributable to urban ambient air pollution. Outdoor air pollution alone causes more than 80,000 hospital admissions per year, nearly 8,000 cases of chronic bronchitis, and almost 5 million cases of lower respiratory cases among children under five. The harm caused by air pollution in Pakistan's urban areas is the highest in the South Asia Region and exceeds most other high profile causes of mortality and morbidity in Pakistan, including traffic-related accidents. Despite the strong evidence indicating an urgent need to improve urban air quality in Pakistan, the issue has received little attention. It is yet to be included as a priority in the country's policy agenda.

Current trends, including industrialization and urbanization, suggest that air quality in Pakistan will worsen over time unless targeted interventions are adopted in the short, medium, and long term, and the institutional and technical capacity of organizations responsible for air quality management (AQM) is strengthened. This book advocates for allocating resources to AQM because there is evidence that AQM is a pillar of green growth. Countries that choose a "clean" development path can grow faster and more sustainably than those that follow a "dirty" trajectory (Esty and Porter 2001; World Bank 2012).

The number of vehicles in Pakistan has jumped from approximately 2 million to 10.6 million over the last 20 years, an average annual growth rate in excess of 8.5%.[4] From 1991 to 2012, the number of motorcycles and scooters grew more than 450%, and motor cars, close to 650% (figure 1.1). The growth rate of mobile sources increased after 2003.

Industrial facilities, particularly those consuming fossil fuels, emit significant amounts of air pollutants. Emissions from large-scale facilities, such as cement, fertilizer, sugar, steel, and power plants—many of which use furnace oil that is high in sulfur content—are a major contributor to poor air quality (Ghauri, Lodhi, and Mansha 2007; Khan 2011). A wide range of small-scale to medium-scale industries, including brick kilns, steel re-rolling, steel recycling, and plastic molding, also contribute significantly to urban air pollution through their use of "waste" fuels, including old tires, paper, wood, and textile waste. The widespread use of small diesel electric generators in commercial and residential areas, which are in response to the electricity outages, further exacerbate industrial emissions. Industrial emissions are associated with poor maintenance of boilers and generators (Colbeck, Zaheer, and Zulfiqar 2010; Ghauri 2010; Ilyas 2007; Khan 2011).[5]

Different nonpoint sources contribute to air pollution in Pakistan, including burning of solid wastes and sugarcane fields. More than 54,000 tons of solid waste are generated daily, most of which is either dumped in low-lying areas or burned. The burning of solid waste at low temperatures produces carbon monoxide (CO), PM, and volatile organic compounds (VOCs), including toxic and carcinogenic pollutants (Faiz 2011). Farmers in Pakistan burn cane fields to ease harvesting. During sugarcane harvesting, high concentrations of particulate matter of less than 10 microns (PM_{10}) are found in rural areas in Punjab and Sindh.

Source apportionment analyses completed in Pakistan (particularly for Lahore) from 2006 to 2012 have found high concentrations of primary and secondary pollutants, particularly fine particulate matter. In Lahore, Stone and others (2010) reported the chemical characterization and source apportionment of fine and coarse PM.[6] The annual average concentration (± one standard deviation) of particulate matter of less than 2.5 microns ($PM_{2.5}$) was 194 ± 94 $\mu g/m^3$. Crustal sources like dust dominated coarse aerosol, whereas carbonaceous aerosol dominated fine particles. While motor vehicle contributions were relatively consistent over the course of the yearlong study, biomass and coal sources demonstrated seasonal variability and peaked in the wintertime. Secondary organic aerosols' contributions also peaked in the wintertime and were potentially enhanced by fog.

In Lahore, dust sources were found to contribute on average 41% of PM_{10} mass and 14% of $PM_{2.5}$ mass on a monthly basis (von Schneidemesser and others 2010). Seasonally, concentrations were found to be lowest during the monsoon season (July–September). Principal component analysis (PCA) identified seven factors: three industrial sources, re-suspended soil, mobile sources, and two regional secondary aerosol sources likely from coal and/or biomass burning. PM measured in Lahore was more than an order-of-magnitude greater

Figure 1.1 Motor Vehicles on the Road (in thousands) in Pakistan, 1991–2012

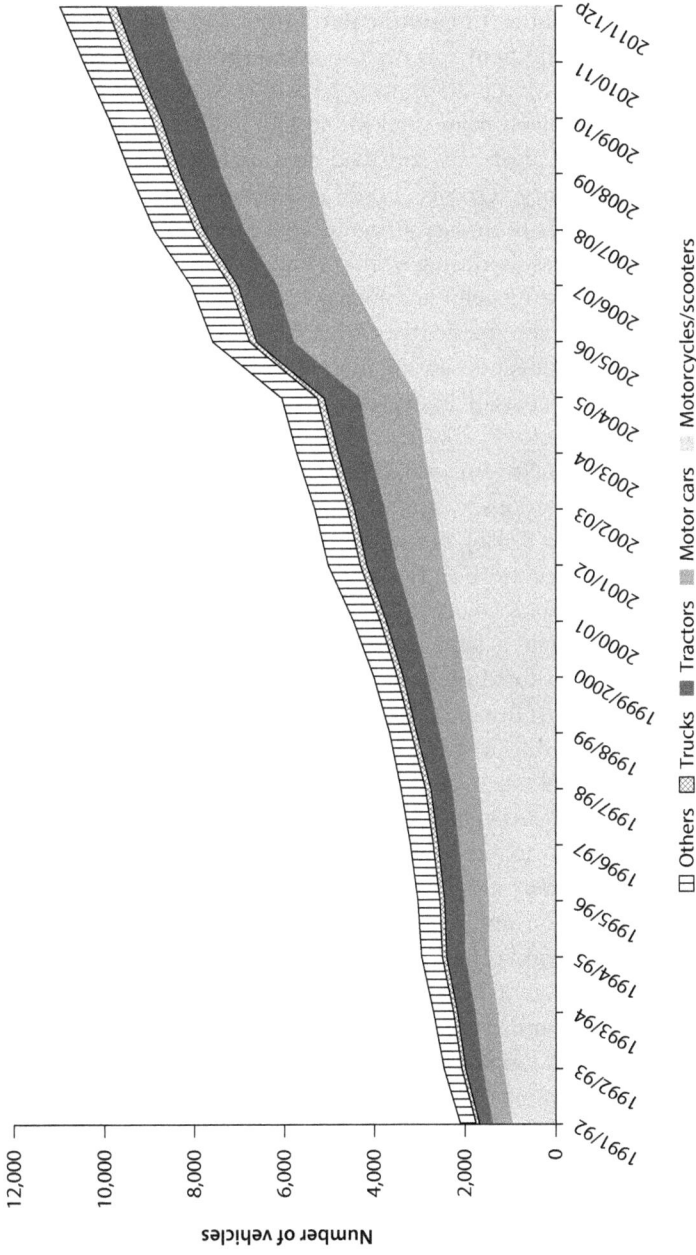

Legend: Others · Trucks · Tractors · Motor cars · Motorcycles/scooters

Source: Ministry of Finance of Pakistan 2012.
Note: Data for 2011–12 are provisional, as indicated by the "p".

than that measured in aerosols from the Long Beach/Los Angeles region and approximately four-fold greater than the activity of Denver area PM (Shafer and others 2010).

Elevated levels of $PM_{2.5}$ at Lahore ranged from 2 to 14 times higher than the prescribed United States Environmental Protection Agency limits (Lodhi 2009). Source apportionment was performed on short duration analysis results of November 2005 to March 2006 using a Positive Matrix Factorization (PMF) model. The results derived from the PMF model indicated that the major contributors to $PM_{2.5}$ in Lahore are soil/road dust, industrial emissions, vehicular emissions, and secondary aerosols (Lodhi and others 2009; Quraishi, Schauer, and Zhang 2009). Trans-boundary air pollutants also affect the city, particularly due to secondary aerosols during winter. The sulfate particles also facilitate haze/fog formation during calm and highly humid conditions, and thus reduce visibility and increase the incidents of respiratory diseases encountered in the city every year. Similar events causing marked reductions in visibility, disrupted transportation, and increased injury and death were also reported in earlier studies (Hameed and others 2000).

Also in Lahore, from November 2005 to January 2006, several sources contributed to $PM_{2.5}$ concentrations including diesel emissions (28%), biomass burning (15%), coal combustion (13%), secondary PM (30%), two-stroke vehicle exhaust (8%), and industrial sources (6%) (Raja and others 2010). Diesel and two-stroke vehicle emissions (36%) accounted for most of the measured high $PM_{2.5}$ mass concentrations. Although a large component of the carbonaceous aerosols in Lahore originated from fossil fuel combustion, a significant fraction was derived from biomass burning (Husain and others 2007). Finally, Zhang, Quraishi, and Schauer (2008), by using a molecular marker-based Chemical Mass Balance (CMB) receptor model, showed that traffic pollution, including exhaust from gasoline- or diesel-powered vehicles, was the predominant source of PM_{10} carbonaceous aerosols. Taken together, gasoline-powered vehicles and diesel exhausts contribute 47.5%, 88.3%, and 15.4% of measured inhalable particulate organic carbon, elemental carbon, and mass, respectively.

In Karachi, Mansha and others (2011) reported a characterization and source apportionment of ambient air particulate matter ($PM_{2.5}$). Source apportionment was performed on PM samples using a PMF model, indicating five major contributors: industrial emissions (53%), soil/road dust (16%), and others that include vehicular emissions (18%), sea salt originating from the Arabian Sea, and secondary aerosols.

In Islamabad, the mean metal concentrations in the atmosphere are far higher than background and European urban sites mainly due to anthropogenic activities (Shah and Shaheen 2009). Industrial metals like iron, zinc, manganese, and potassium showed viable correlations while Pb is correlated with cadmium because of their common source. Principal component and cluster analyses revealed automobile emissions, industrial activities, combustion processes, and mineral dust as the major pollution sources in the atmospheric particles (Shah and Shaheen 2007; Shah, Shaheen, and Nazir 2012). The comparison study presents high concentrations of airborne trace metal. The basic statistical data revealed quite divergent variations of the elements during the specific seasons (Shah and Shaheen 2010). Automobile emissions, windblown soil dust, excavation activities, biomass burning, and industrial and fugitive emissions were identified as major pollution sources in the atmospheric aerosols of Islamabad. The comparative data showed that the concentrations of airborne trace elements in this area are mostly very high compared with other regions, thus posing a potential health hazard to the local population.

Current trends in Pakistan suggest that, in the absence of targeted interventions, urban air pollution could become even more severe in the future and affect a larger number of people. Services represented 52.7% of Pakistan's gross domestic product (GDP) in 2008 and grew at average annual rate of 5.7% during 1998–2008. In comparison, industry's share was 26.9% of GDP and grew at an average rate of 6.6% during the same period, while agriculture amounted to 20.4% of GDP and grew at a yearly average rate of 3.1%. Current projections indicate that services will remain the economy's most important sector in the future and that industry's contribution will increase as agriculture's share falls. These changes in the economy's structure are likely to be associated with increased urbanization. Pakistan is already the most urbanized large country in South Asia, with around 37% of its population living in cities in 2010, and it is estimated that more than 60% of the country's population will be urban by 2050.

Industrialization and urbanization, in conjunction with motorization, could result in further deterioration of urban air quality. Industries tend to agglomerate in a few geographic locations where there is availability of specialized labor, inter-industry spillovers, higher road density, local transfer of knowledge, and access to international supplier and buyer networks. These factors enhance firm competitiveness and largely explain the clustering of large-scale manufacturing and high employment levels around the metropolitan areas of Karachi and Lahore. While agglomeration economies can generate significant benefits, they can also result in congestion and air pollution, among other public "bads." Given the already severe damages caused by air pollution and the possibility that environmental conditions might worsen because of growing industrialization and urbanization, the Government of Pakistan (GoP) should consider adopting priority interventions in the short term and building the institutional and technical capacity to adopt additional measures over the medium and long term.

This book examines policy options to strengthen the Pakistan Clean Air Program (PCAP) to better address the cost imposed by outdoor air pollution upon Pakistan's economy and populace. Few interventions will provide returns as high as those in air pollution control in Pakistan will. The proposed approach focuses on strengthening institutions to carry out priority interventions on ambient AQM through learning by doing.

The approach proposed in this book recommends that the federal and provincial environmental protection agencies (EPAs) take on a limited number of high-return, essential, and feasible interventions drawn largely from the PCAP.

The planning and implementation of the menu of policy and management options provides the basis for growing results-based management and good governance. It is also expected to measurably increase ambient air quality and reduce health damage and associated air pollution costs. Experiences from around the world indicate that the proposed interventions can significantly improve urban air quality, if they receive strong political commitment to ensure their timely and sustained implementation.

Objective

The objective of this book is to examine policy options to control outdoor air pollution in Pakistan. The findings of this analysis aim at assisting the GoP in the design and implementation of reforms to improve the effectiveness and efficiency of Pakistan's ambient air quality institutions. The overarching theme of this book is that prioritizing interventions is essential to address the cost of outdoor air pollution, given current resource limitations. These priority interventions are strategic, cost effective, and proven somewhere else in Asia. Most of these interventions involve strengthening the institutional capacity of Pakistan's preeminent environmental regulatory agencies, the Pakistan Environmental Protection Agency (Pak-EPA), provincial EPAs, and others.

Methodology

Methodology for Air Quality Data Analysis

The provincial EPAs and the Pak-EPA are in charge of monitoring air pollution in Pakistan. From 2006 to 2009, the Japanese International Cooperation Agency (JICA) assisted the GoP in the design and installation of an air quality monitoring network of measurement stations that included (a) fixed and mobile air monitoring stations in five major cities of Pakistan (Islamabad, Karachi, Lahore, Peshawar, and Quetta); (b) a data center; and (c) a central laboratory. The monitoring units in the provinces are managed and operated by the provincial EPAs. Actual operation was initially carried out by consultants hired, trained, and paid by the Japanese partners, and after the first year of operation, the EPAs were expected to assume all costs related to the monitoring work.

Administrative and budget problems have led to inadequate operation and maintenance of the air quality monitoring network. In mid-2012, the monitoring network suspended its operations because of the EPAs' failure to cover its operations and maintenance costs. Automated instruments were used in Lahore, Karachi, Quetta, Peshawar, and Islamabad, while manual, or manual and automated combinations were used in other areas. The number and location of the monitors were criticized because they exclude "hot spots" where air quality was believed to be particularly poor. Various technical problems were reported related to the interrupted power supply and difficulties in maintaining the automated electronic instruments. Furthermore, the data that were collected were neither analyzed nor disclosed, and the concentration of $PM_{2.5}$ was being monitored infrequently. As emphasized in this book despite the installation of this air quality monitoring data, the reliability of air quality data are suboptimal in Pakistan. Furthermore, air quality data collected before 2004 are significantly less reliable than the data collected by this air quality monitoring network.

While there is a paucity of available air quality data for Pakistan, JICA's funded air quality monitoring network collected information on concentrations of PM, carbon monoxide, sulfur and nitrogen oxides, ozone, and other parameters in Pakistan's major cities.[7] However, daily average ambient air quality data are available for less than 50 percent of the days between July 26, 2007, and April 27, 2010. Table 1.1 shows the temporal (percent of days) coverage for all the air quality data and points to the lowest coverage for $PM_{2.5}$.

Despite its flaws, these data provide the best picture of air quality in Pakistan. Using these data, average values were calculated for all air quality pollutants. PCA was performed using the STATISTICA v4.2 software on a dataset composed of meteorological parameters, gaseous pollutants, and $PM_{2.5}$ mass. This methodology combines a factor analysis that results in the identification of potential pollution sources, indicates the seasonal evolution of the sources, and quantifies the annual mean contribution for each one.

Methodology for Economic Analysis

The environmental health and economic analysis relies on primary data obtained from various ministries, agencies, and institutions in Pakistan, as well as from international development agencies. The analysis also uses several hundred reports and research studies from Pakistan and other countries. Quantification of health effects from environmental risk factors is grounded in commonly used methodologies that link health outcomes and exposure to pollution and other health risk factors, and the economic costs of these health effects are estimated using standard valuation techniques. The assessment of the benefits and costs of interventions to mitigate health effects and improve natural resource conditions is based on these same methodologies and valuation techniques, as well as on international evidence of intervention effectiveness, and, to the extent available, on data regarding the costs of interventions in Pakistan.

Table 1.1 Data Coverage (Percentage) for Air Quality Parameters for Five Cities in Pakistan

Data coverage (%)	$PM_{2.5}$	SO_2	NO_2	O_3	CO
Islamabad	42	31	78	87	96
Quetta	5	46	53	29	53
Karachi	10	23	29	24	24
Peshawar	17	35	66	28	60
Lahore	12	54	75	79	44
Average	*17*	*38*	*60*	*49*	*55*

Source: Estimates by Dall'Osto 2012.
Note: $PM_{2.5}$ = particulate matter of less than 2.5 microns , SO_2 = sulfur dioxide, NO_2 = nitrogen dioxide, O_3 = ozone, CO = carbon monoxide.

Methodology for Institutional Analysis

Pakistan has institutional foundations that could support the development of priority actions targeting the country's severe air pollution problem. However, a number of obstacles have hampered the development of adequate responses, including acute institutional weaknesses and lack of funding. The institutional analysis included in this book assessed the legal mandates of environmental organizations at the federal and provincial level, particularly after the 18th Constitutional Amendment. Based on those mandates, the analysis evaluates the adequacy of human resources to carry out key technical, management, and support functions, as well as the availability of physical capital needed to perform the assigned functions. The institutional analysis includes a review of the existing formal rules (for example, laws, policies, and standards), informal rules (for example, the capacity of powerful groups to reverse the adoption of pollution charges, and the pervasive tolerance of violators of environmental regulations) and their enforcement mechanisms. In addition, the analysis looks at the role that the courts have played and could potentially play in the future when environmental organizations fail to fulfill their AQM responsibilities.

The analysis presented in this book uses primary data collected from air quality monitoring stations installed by JICA in 2006. After developing an air quality baseline for Pakistan (Ghauri, Lodhi, and Mansha 2007), the World Bank contracted a number of specialists to complete an analysis of alternatives to strengthen Pakistan AQM institutions. The book's scope, methodology, and findings benefited from significant stakeholder input. Preliminary findings from this analysis were discussed at the first AQM consultation workshop on January 13, 2010, with government representatives and other stakeholders. Both the stakeholders at this workshop and the Pak-EPA recommended that the scope of the analytical work be increased to include a discussion of international examples and a discussion of climate-change co-benefits. Their comments were incorporated into an expanded draft, which was discussed at the Second AQM Consultation Workshop. The stakeholders who attended the workshop included representatives from the Ministry of Environment (MoE),[8] the Planning Commission, provincial governments, sectoral ministries, and development partners. The book has also counted on the close participation of the Pak-EPA, which provided technical supervision and timely comments during the whole process.

The book also includes a review of secondary sources, focusing on recent analysis of the effects of different air pollutants on human health, as well as lessons learned from ongoing regional and international efforts to improve ambient air quality. Since the 18th Constitutional Amendment devolved responsibilities for environmental management to the provinces, the government requested the task team to include in the book a discussion on approaches to decentralization, using references to international examples. In response to this request, the book examines the lessons from the decentralization of environmental management functions in several developed and developing countries. The book indicates which functions have been found to be managed more effectively at the central level and which can be devolved to subnational levels to achieve better results. In addition, the book discusses the need for an apex body, even in highly decentralized systems.

Analytical Value-Added

At the request of the Pak-EPA, this book has analyzed a number of topics in AQM that had not previously been discussed regarding Pakistan. These elements include the following:

- Additional primary data collection on ambient air quality levels in Pakistan's major cities
- An analysis of the effect of fuel price distortions on Pakistan's air quality and benefit and cost estimates for alternatives for reducing the sulfur content of fuel in Pakistan
- Estimation of the costs and benefits of various interventions to decrease emissions from mobile and stationary sources
- An analysis of emissions levels and possible air pollution control interventions for Pakistan's industrial sector and other stationary and diffuse sources
- Discussion of the type and likelihood of possible climate-change mitigation-related benefits from interventions to improve air quality in Pakistan
- An institutional analysis of environmental management agencies in Pakistan that discusses possible approaches to decentralization, using references to international examples.

Structure of the Book

This book has seven chapters. Chapter 2 identifies major trends in ambient air pollution, including concentration levels of main pollutants and the identification of principal sources. The chapter shows that air pollution is one of the most widespread and serious problems in Pakistan's cities and provides estimates of the significant damages that urban air pollution causes every year to the country's population and economy. The chapter also summarizes the available data on air quality in Pakistan's urban centers, which indicate that pollution concentrations significantly exceed limits considered safe for human beings.

Chapter 3 examines the evolution of Pakistan's AQM framework over the period 1993–2013. The organizational roles defined by the Pakistan Environmental Protection Act (PEPA) include financially independent provincial environmental agencies, a central-level division responsible for overall policy formulation and coordination, and a central enforcement agency. After 2011, provincial governments took over environmental management responsibilities from the provinces. This chapter discusses policy options for strengthening Pakistan's EPAs, taking into account the devolution of environmental management responsibilities. This chapter also provides an assessment of Pakistan's capacity to monitor outdoor air pollution in urban areas.

Cleaning Pakistan's Air • http://dx.doi.org/10.1596/978-1-4648-0235-5

Chapter 4 examines options to control air pollution from mobile sources, the main contributors of several air pollutants, including noxious fine particulate matter ($PM_{2.5}$) and its precursors. The chapter evaluates the costs and benefits of vehicle inspection and scrappage programs, which could initially be implemented in Pakistan's largest urban areas, where air pollution is particularly acute. It discusses alternatives for improving diesel quality, including importing low-sulfur diesel and retrofitting of refineries. A summary on efforts to convert to compressed natural gas (CNG) vehicles and challenges ahead is also included. Finally, the chapter looks at the effects of fuel price distortions and how their phaseout would contribute to improve urban air quality.

Chapter 5 addresses measures to tackle pollution from industrial sources. This chapter offers an analysis of alternatives to control air emissions from industrial and other stationary sources, including economic instruments, command and control regulations, legal actions, public disclosure mechanisms, and cleaner production efforts. The chapter highlights the importance of establishing an inventory of stationary air pollution sources. In addition, the chapter looks at the contributions of nonpoint sources, particularly burning of municipal wastes and agricultural residues, and policy options to control them.

Chapter 6 identifies synergies of interventions for air pollution control and climate change mitigation. The chapter explains the mutually reinforcing interactions between climate change and ambient air pollution. The chapter discusses the contributions of the energy, transportation, and agricultural sectors to greenhouse gas emissions in Pakistan. It also outlines the policies that could be implemented in these sectors and that have the potential to deliver both climate change and local air quality benefits. The chapter describes the potential role voluntary and compliance carbon markets could have in supporting Pakistan's climate change and local air quality efforts.

Chapter 7 summarizes the main conclusions of the book. The chapter highlights potential policy and institutional reforms needed to address AQM priorities and to target complementary investments. The chapter integrates a limited number of high-return, essential, and feasible interventions that the GoP can implement to address the severe health and economic damages caused by ambient air pollution. The proposed interventions are largely drawn from PCAP, are informed by regional and international best practices, and focus on strengthening existing institutions responsible for ambient AQM through learning by doing.

Notes

1. Estimates from the Global Model of Ambient Particulates indicate that as of 2002, average PM_{10} in Pakistan was 165 µg/m³ (World Bank 2006). These estimates were based on an econometrically estimated model for predicting PM levels.

2. Gurjar and others (2008) suggested a multi-pollutant index that takes into account the combined level of the three WHO criteria pollutants and found that Karachi, one of the mega-cities of Pakistan, held the fourth position worldwide on the multi-pollutant index-based ranking.

3. Recent advances in remote sensing have provided new avenues for measuring, monitoring, and understanding processes that lead to atmospheric pollution. Mansha and Ghauri (2011) assessed a seasonal and spatial variation of aerosol concentration over Karachi, Pakistan, using satellite-derived aerosol optical depth (AOD) data as well as other ground-based data, including $PM_{2.5}$. It was demonstrated that satellite information can be used with a reasonable degree of accuracy, particularly when limited ground-based measurements are available. Recently, Alam and others (2010) analyzed the spatio-temporal variation of AOD and the aerosol influence on cloud parameters using moderate resolution imaging spectroradiometer data and found higher AOD values in summer and lower values in winter for various cities in Pakistan. Alam, Trautmann, and Blaschke (2011) studied the aerosol optical properties and radiative forcing over mega-city Karachi. The high July AOD values are predominantly caused by dust activities in the southern part of Pakistan, whereas high AOD values for November are due to high black carbon concentrations and to some dust plumes near Karachi (Dutkiewicz and others 2009). Alam and others (2012) again reported high AOD values over Lahore and Karachi during December, largely due to absorbing anthropogenic activities, while high AOD values in June were attributed to mineral dust aerosols. Alam and others (2011) used three different satellite-borne sensors to investigate the spatial and temporal variations of aerosols over several cities in Pakistan. Back trajectory analyses indicated that while winter air masses reaching Pakistan had traveled long distances, summer air masses had traveled only short distances. While monsoonal rainfall tends to reduce PM concentrations by washing aerosols out of the atmosphere, this effect was found to be mainly restricted to the eastern and southeastern parts of Pakistan.

4. The figures refer to motor vehicles on the road, which according to the Ministry of Finance (2010), are significantly more than the number of registered vehicles.

5. The results reported by Khan (2011) and Ghauri (2010) are part of a World Bank-financed study that present the data obtained by the consultants through a survey and interviews among different industrial units and other stakeholders in Pakistan.

6. Ambient air quality in Lahore has documented 24-h maximum wintertime $PM_{2.5}$ concentrations of 200 $\mu g/m^3$ (Biswas, Ghauri, and Husain 2008; Zhang, Auraishi, and Schauer 2008). Principal component analysis (PCA) showed that the greatest contributors to PM_{10} were industrial sources, followed by secondary aerosols and mobile sources.

7. The data on emissions sources and inventories in Pakistan need to be improved to refine the design and targeting of control strategies for both mobile and stationary sources.

8. After the adoption of the 18th Constitutional Amendment, the Ministry of Environment was subsumed by the Ministry of National Disaster Management, which became the Ministry of Climate Change in April 2012. However, the Ministry of Climate Change was downgraded to a division in June 2013.

Bibliography

Alam, K., T. Blaschke, P. Madl, A. Mukhtar, M. Hussain, T. Trautmann, and S. Rahman. 2011. "Aerosol Size Distribution and Mass Concentration Measurements in Various Cities of Pakistan." *Journal of Environmental Monitoring* 13: 1944–52.

Alam, K., M. J. Iqbal, T. Blaschke, S. Qureshi, and G. Khan. 2010. "Monitoring Spatio-Temporal Variations in Aerosols and Aerosol-Cloud Interactions over Pakistan Using MODIS Data." *Advances in Space Research* 46: 1162–76.

Alam, K., S. Qureshi, and T. Blaschke. 2011. "Monitoring Spatio-Temporal Aerosol Patterns Over Pakistan Based on MODIS, TOMS and MISR Satellite Data and a HYSPLIT Model." *Atmospheric Environment* 45: 4641–51.

Alam, K., T. Trautmann, and T. Blaschke. 2011. "Aerosol Optical Properties and Radiative Forcing Over Mega City Karachi." *Atmospheric Research* 101: 773–82.

Alam, K., T. Trautmann, T. Blaschke, and H. Majid. 2012. "Aerosol Optical and Radiative Properties during Summer and Winter Seasons over Lahore and Karachi." *Atmospheric Environment* 50 (April), 234–45. doi: 10.1016/j.atmosenv.2011.12.027.

Aziz, A., and I. U. Bajwa. 2004. "Energy and Pollution Control Opportunities for Lahore." *Urban Transport X* 16: 751–60.

Aziz, J. A. 2006. "Towards Establishing Air Quality Guidelines for Pakistan." *East Mediterranean Health Journal* 12 (6): 886–93.

Biswas, K. F., B. M. Ghauri, and L. Husain. 2008. "Gaseous and Aerosol Pollutants during Fog and Clear Episodes in South Asian Urban Atmosphere." *Atmospheric Environment* 42 (33): 7775–85.

Colbeck, I., A. Zaheer, and A. Zulfiqar. 2010. "The State of Ambient Air Quality in Pakistan: A Review." *Environmental Science and Pollution Research* 17: 49–63. http://www.springerlink.com/content/f718jn535422j0wh/fulltext.pdf.

Dall'Osto, M. 2012. Unpublished consultant report prepared for the World Bank, Washington, DC.

Dutkiewicz, V. A., S. Alvi, B. M. Ghauri, M. I. Choudhary, and L. Hussain. 2009. "Black Carbon Aerosols in Urban Air in South Asia." *Atmospheric Environment* 43: 1737–44.

Esty, D., and M. E. Porter. 2001. "Ranking National Environmental Regulation and Performance: A Leading Indicator of Future Competitiveness?" *Global Competitiveness Report 2001–2002.* New York: Oxford University Press. http://www.stadt-zuerich.ch /content/dam/stzh/prd/Deutsch/Stadtentwicklung/Publikationen_und_Broschueren /Wirtschaftsfoerderung/Standort_Zuerich/GCR_20012002_Environment.pdf.

Faiz, A. 2011. *Comments and Edits on Draft Air Quality Management Report.* E-mail messages from June 19, 26, 29, and 30. Personal communication with authors.

Ghauri, B. 2010. *Institutional Analysis of Air Quality Management in Urban Pakistan.* Study commissioned by the World Bank, Washington, DC. cleanairinitiative.org/portal /system/files/.../AQM_Draft_Final_Report.pdf.

Ghauri, B., A. Lodhi, and M. Mansha. 2007. "Development of Baseline (Air Quality) Data in Pakistan." *Environmental Monitoring and Assessment* 127 (1–3): 237–52. http://www.springerlink.com/content/ak3g134828122567/fulltext.pdf.

Gurjar, B. R., T. M. Butler, M. G. Lawrence, and J. Lelieveld. 2008. "Evaluation of Emissions and Air Quality in Megacities." *Atmospheric Environment* 42: 1593–1606.

Hameed, S., M. I. Mirza, B. M. Ghauri, Z. R. Siddiqui, R. Javed, A. R. Khan, O. V. Rattigan, S. Qureshi, and L. Husain. 2000. "On the Widespread Winter Fog in Northeastern Pakistan and India." *Geophysical Research Letters* 27 (13): 1891–94.

Husain, L., V. A. Dutkiewicz, A. J. Khan, and B. M. Ghauri. 2007. "Characterization of Carbonaceous Aerosols in Urban Air." *Atmospheric Environment* 4 (32): 6872–83.

Ilyas, S. Z. 2007. "A Review of Transport and Urban Air Pollution in Pakistan." *Journal of Applied Sciences and Environmental Management* 11 (2): 113–21. http://www.ajol .info/index.php/jasem/article/viewFile/55004/43484.

Khan, A. U. 2011. *Industrial Environmental Management in Pakistan*. January 2011. Consultant report prepared for the World Bank, Washington, DC.

Lodhi, A., G. Badar, M. K. Rafiq, S. Rahman, and S. Shoaib. 2009. "Particulate Matter ($PM_{2.5}$) Concentration and Source Apportionment in Lahore." *Journal of the Brazilian Chemical Society* 20 (10): 1811–20.

Lodhi, Z. H. 2009. "Ambient Air Quality in Pakistan." Pak-EPA (accessed October 20, 2009), http://www.environment.gov.pk/PRO_PDF/AmbientAirQtyPakistan.pdf.

Mansha, M., and B. Ghauri. 2011. "Assessment of Fine Particulate Matter ($PM_{2.5}$) in Metropolitan Karachi Through Satellite and Ground-based Measurements." *Journal of Applied Remote Sensing* 5. doi: 10.1117/1.3625615.

Mansha, M., B. Ghauri, S. Rahman, and A. Amman. 2011. "Characterization and Source Apportionment of Ambient Air Particulate Matter ($PM_{2.5}$) in Karachi." *Science of the Total Environment*. doi: 10.1016/j.scitotenv.2011.10.056.

Ministry of Finance. 2010. *Pakistan Economic Survey 2009–10*. Islamabad: Ministry of Finance, Government of Pakistan. http://www.finance.gov.pk/survey_0910.html.

———. 2012. *Pakistan Economic Survey 2011–12*. Islamabad: Ministry of Finance, Government of Pakistan. http://www.finance.gov.pk/survey_1112.html.

Quraishi, T. A., J. J. Schauer, and Y. Zhang. 2009. "Understanding Sources of Airborne Water Soluble Metals in Lahore, Pakistan." *Kuwait Journal of Science* 36 (1): 43–62.

Raja, S., K. F. Biswas, L. Hisain, and P. K. Hopke. 2010. "Source Apportionment of the Atmospheric Aerosol in Lahore, Pakistan." *Water, Air & Soil Pollution* 208 (1–4): 43–57.

Shafer, M. M., D. A. Perkins, D. S. Antkiewicz, E. A. Stone, T. A. Quraishi, and J. J Schauer. 2010. "Reactive Oxygen Species Activity and Chemical Speciation of Size-fractionated Atmospheric Particulate Matter from Lahore, Pakistan: An Important Role for Transition Metals." *Journal of Environmental Monitoring* 12 (3): 704–15.

Shah, M. H., and N. Shaheen. 2007. "Statistical Analysis of Atmospheric Trace Metals and Particulate Fractions in Islamabad, Pakistan." *Journal of Hazardous Material* (Elsevier), 147 (3): 759–67. doi: 10.1016/j.jhazmat.2007.01.075.

———. 2009. "Study of Particle Size and Trace Metal Distribution in Atmospheric Aerosols of Islamabad." *Journal of the Chemical Society of Pakistan* 31 (3): 427–33.

———. 2010. "Seasonal Behaviors in Elemental Composition of Atmospheric Aerosols Collected in Islamabad, Pakistan." *Atmospheric Research* 95 (2–3): 210–23. doi: 10.1016/j.atmosres.2009.10.001.

Shah, M. H., N. Shaheen, and R. Nazir. 2012. "Assessment of the Trace Elements Level in Urban Atmospheric Particulate Matter and Source Apportionment in Islamabad, Pakistan." *Atmospheric Pollution Research*. doi: 10.5094/APR.2012.003.

Stone, E., J. Schauer, T. A. Quraishi, and A. Mahmood. 2010. "Chemical Characterization and Source Apportionment of Fine and Coarse Particulate Matter in Lahore, Pakistan." *Atmospheric Environment* 44: 1062–70.

von Schneidemesser, E., E. A. Stone, T. Quraishi, M. M. Shafer, and J. J. Schauer. 2010. "Toxic Metals in the Atmosphere in Lahore, Pakistan." *Science of the Total Environment* 408 (7): 1640–48.

World Bank. 2006. "Pakistan Strategic Country Environmental Assessment." South Asia Environment and Social Development Unit, Washington, DC. http://www.esmap.org

/esmap/sites/esmap.org/files/FR275-03_Thailand_Reducing_Emissions_from
_Motorcycles_in_Bangkok.pdf.

———. 2012. *Inclusive Green Growth. The Pathway to Sustainable Development.*
Washington, DC: World Bank.

Zhang, Y., T. Quraishi, and J. J. Schauer. 2008. "Daily Variations in Sources of Carbonaceous
Aerosol in Lahore, Pakistan during a High Pollution Spring Episode." *Aerosol and Air
Quality Research* 8 (2): 130–46.

CHAPTER 2

Air Pollution in Pakistan

Introduction

In the South Asia Region, Pakistan ranks as the worst in air pollution measured as particulate matter (PM) (measures PM_{10}).[1] Concentrations of noxious particulate matter (PM) in Pakistan are significantly higher than those found across South Asia (figure 2.1).

Air pollutants that cause negative health and environmental impacts include PM, ozone (O_3), carbon monoxide (CO), nitrogen oxides (NO_x), sulfur oxides (SO_x), and volatile organic compounds (VOCs), as described below.

- PM is the term for airborne particles; it includes dust, dirt, soot, smoke, and liquid droplets. It is the air pollutant most damaging to health. Some particles are directly emitted into the air. Particles can also be created by atmospheric conversion of sulfur dioxide (SO_2) and NO_x into sulfates and nitrates. Monitoring stations that measure PM_{10} are located only in Islamabad, Quetta, Peshawar, Lahore, and Karachi. PM_{10} can be inhaled into the lungs and results in respiratory illness associated with premature mortality. Despite strong scientific evidence that particulate matter of less than 2.5 microns ($PM_{2.5}$) is the most dangerous subset of PM_{10}, no systematic monitoring information on $PM_{2.5}$ is available in Pakistan.[2]
- SO_2 is a by-product of burning fossil fuels such as crude oil, furnace oil, diesel, and coal. SO_2 can be transformed in the atmosphere into sulfates that appear as fine particles.
- NO_x derive from vehicle exhaust, combustion installations such as power plants, and industrial and agricultural activity. NO_x react with other air pollutants to form O_3 and fine particulates (nitrates) in the lower atmosphere.
- O_3 VOCs react with NO_x and other chemicals in the atmosphere to create harmful secondary pollutants, including O_3, and cause health problems ranging from eye irritation to decreased lung capacity.
- CO is a product of incomplete fuels combustion. At low exposure levels, CO causes mild effects that include headaches, dizziness, disorientation, nausea, and fatigue. However, high exposure levels of CO might be lethal.

Figure 2.1 PM$_{10}$ Concentrations at the Country Level (µg/m³), 2008

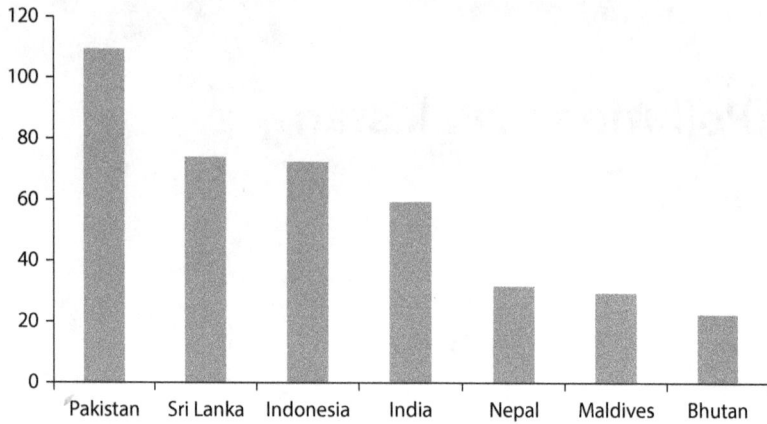

Source: World Bank 2010a.

- VOCs encompass various contaminants, including organic compounds and solvents. Petroleum reservoirs, storage systems for gasoline and other volatile compounds, industrial processes and fuel combustion, use of paints and cleaners, and agricultural activities emit these compounds. VOCs are the main precursor of O_3.

The most important economic costs associated with air pollution correspond to impacts on human health. Health impacts of, for example, urban air pollution are expressed in terms of (a) excess mortality; and (b) loss of disability-adjusted life years (DALYs), including mortality loss. Human health effects made up 94% of total economic damages associated with air pollution, out of which premature deaths account for 71% of global annual damages and morbidity accounts for 23% of global annual damages (Muller and Mendelsohn 2007).

Research carried out from 1992 to 2010 indicates that the most severe economic damages from air pollution originate from the exposure to PM$_{2.5}$ and O_3 (Pope and Dockery 2006; Pope and others 2002; WHO 2006). However, health impacts of O_3 appear to be less significant than those associated with exposure to PM. In South Asia and globally, concentrations of O_3 have less impact on human health than those of fine particulate matter PM$_{2.5}$ (table 2.1) (OECD 2008; WHO 2006).

Air quality data published by Ghauri, Lodhi, and Mansha (2007) show that urban concentrations of PM$_{10}$ in Pakistan frequently exceed 200 micrograms per cubic meter (µg/m³). Such concentrations are substantially larger than the World Health Organization (WHO) interim target of 70 µg/m³. In contrast, data compiled by Colbeck, Zaheer, and Zulfiqar (2010) demonstrate that concentrations of O_3 in Pakistan's large cities have been well within the WHO air quality guidelines. Colbeck, Zaheer, and Zulfiqar (2010) also indicate that it is very likely that CO levels in urban centers exceed the WHO guidelines during the day, particularly

Quetta (with a 48-hour mean of 8.1 mg/m^3), Karachi (5.8 mg/m^3), Rawalpindi, Lahore (4.6 mg/m^3), and Islamabad and Peshawar (3.5 mg/m^3). However, the damages caused by CO are not as significant as those of PM$_{2.5}$. This chapter examines the concentrations of air pollutants in Pakistan and analyzes the cost of environmental degradation associated with outdoor air pollution. The chapter identifies PM$_{2.5}$ as the most damaging air pollutant in Pakistan, and estimates the cost of ambient air pollution in Pakistan, particularly in Karachi.

Analysis of Air Quality Data

Temporal Trends and Source Apportionment of Air Pollutants

An analysis of temporal trends of air quality pollutants for Islamabad, Karachi, Peshawar, and Lahore shows higher CO, NO$_2$, SO$_2$, and PM$_{2.5}$ concentrations during winter periods; O$_3$ shows the opposite trend. Since sunlight and heat drive the O$_3$ formation, warm sunny days usually have more O$_3$ than cool or cloudy days (table 2.3). All pollutants present similar wind roses (figure 2.2).

Table 2.1 Premature Deaths Associated with PM and O$_3$ per Million Inhabitants

Region	Premature deaths per million inhabitants—PM	Premature deaths per million inhabitants—O$_3$
Pacific	20	1
Europe	71	8
North America	79	16
Asia	123	14
Brazil	85	3
Russian Federation	149	15
China	266	8
South Asia (including India)	175	5
Rest of the World	129	3
World	158	7

Source: OECD 2008.
Note: PM = particulate matter, O$_3$ = ozone.

Table 2.2 Average Meteorological Values at Four Selected Cities: Peshawar (PES), Lahore (LAH), Karachi (KAR), and Islamabad (ISL)

City	WS (m/s)	Cloud base (altitude, m)	Cloudiness (%)	Pressure (mm [hPa])	RH (%)	Temp (°C)	Visibility (km)
ISL	4.3 ± 2.3 (0–18)	2,500 ± 580 (450–3,000)	2.3 ± 1.5 (0–9)	840 ± 30 (750–890)	48 ± 12 (21–93)	17 ± 9 (– 4.8/35)	6 ± 2 (0.2–11.5)
KAR	2.6 ± 1.6 (0–14)	2,080 ± 900 (800–3,000)	3.7 ± 2.2 (0–10)	1,005 ± 40 (780–1,050)	60 ± 15 (19–93)	27 ± 4 (11–35)	4 ± 1 (2–15)
PES	2.9 ± 1.9 (0–14)	1,980 ± 700 (450–3,000)	2.9 ± 1.5 (0–9)	960 ± 33 (880–990)	61 ± 13 (17–100)	23 ± 8 (6–42)	4 ± 1 (0–10)
LAH	1.1 ± 2 (0–12)	2,600 ± 700 (1,025–3,000)	2.6 ± 1.5 (960–1,000)	980 ± 33 (960–1,000)	60 ± 15 (20–97)	25 ± 7 (7–38)	4 ± 1 (1–7)

Source: Dall'Osto 2012.
Note: WS = wind speed; RH = relative humidity.

Figure 2.2 Wind Roses for PM$_{2.5}$, SO$_2$, NO$_2$, O$_3$, and CO for Four Different Cities: Peshawar (PES), Lahore (LAH), Karachi (KAR), and Islamabad (ISL)

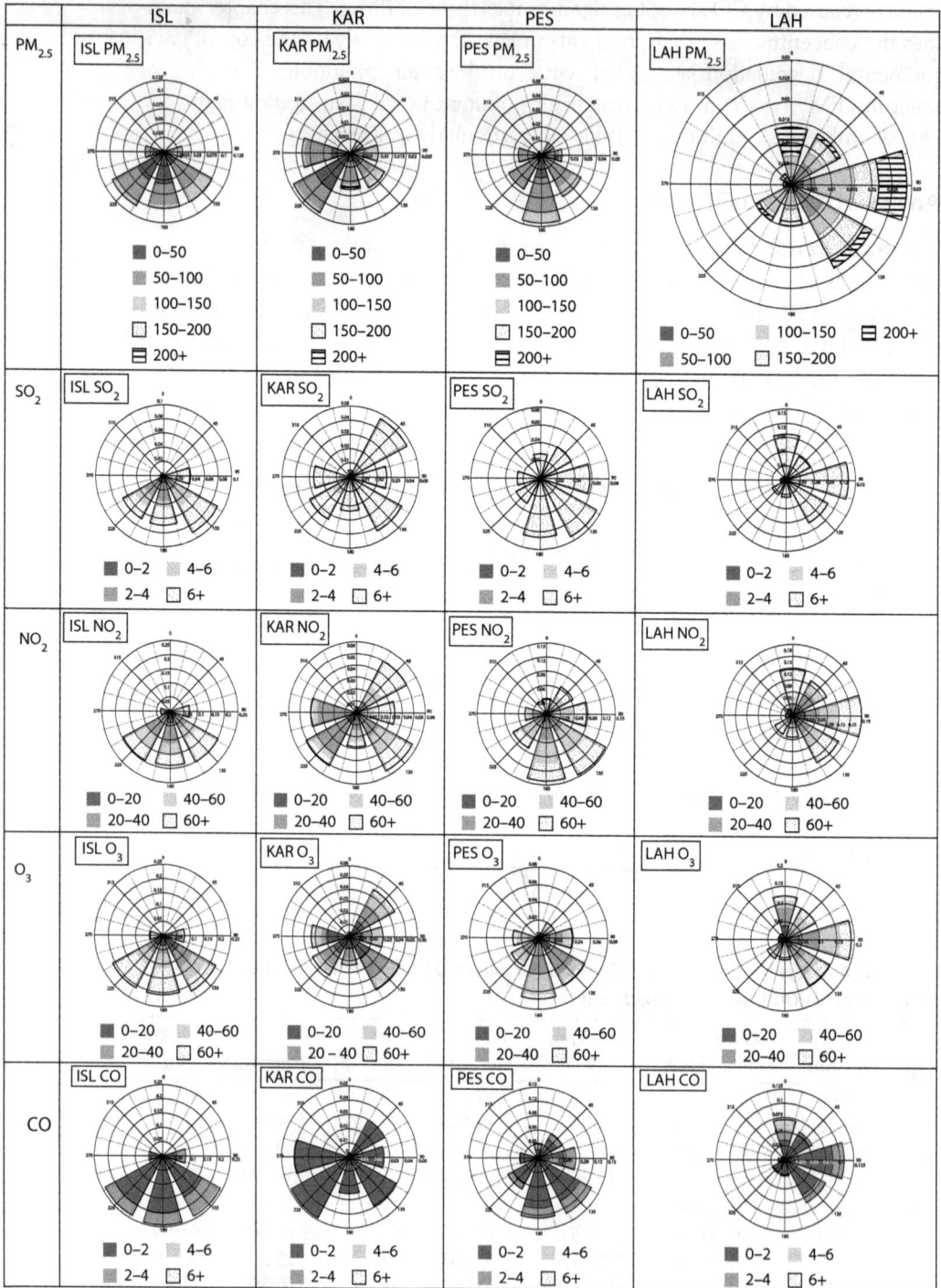

Source: Dall'Osto 2012.

Note: PM$_{2.5}$ = particulate matter of less than 2.5 microns, SO$_2$ = sulfur dioxide, NO$_2$ = nitrogen dioxide, O$_3$ = ozone, CO = carbon monoxide.

During winter periods, SO_2 and $PM_{2.5}$ tend to present higher concentrations; O_3 shows the opposite trend.

Thermal inversions that happen across much of Pakistan from December to March lower the mixing height and result in high pollutant concentrations, especially under stable atmospheric conditions. Suspended PM contributes to the formation of ground fog that prevails over much of the Indo-Gangetic Plains during the winter months. In addition, sunny and stable weather conditions lead to high concentration of pollutants in the atmosphere (Sami, Waseem, and Akbar 2006). Due to high temperatures in summer (40–50°C), fine dust rises up with the hot air and forms "dust clouds" and haze over many cities of southern Punjab and upper Sindh. Dust storms generated from deserts (Thal, Cholistan, and Thar), particularly during the summer, adversely affect visibility in the cities of Punjab and Sindh.

Concentrations of Air Pollutants

An analysis of the available data from 2007 to 2010 shows very high concentrations of fine particle matter ($PM_{2.5}$) in Lahore (143 µg/m³), Peshawar (71 µg/m³), Karachi (88 µg/m³), Islamabad (61 µg/m³), and Quetta (49 µg/m³) (table 2.3). The high-value concentrations reported in this analysis are likely to be even higher if monitoring instruments had been working all the time. Particulate matter of less than 1 micron (PM_1) and PM_{10} measurements were not available, and low data coverage (average 17%) partially affected $PM_{2.5}$ measurements (table 2.4).

Table 2.3 Average Values of $PM_{2.5}$, SO_2, NO_2, O_3, and CO for Five Different Cities: Peshawar (PES), Lahore (LAH), Karachi (KAR), Islamabad (ISL), and Quetta (QUE)

Average	$PM_{2.5}$ (µg/m³)	SO_2 (µg/m³)	NO_2 (µg/m³)	O_3 (µg/m³)	CO (mg/m³)
ISL	61 ± 31	6 ± 3	49 ± 28	47 ± 29	1.4 ± 1
QUE	49 ± 26	54 ± 26	37 ± 15	40 ± 13	1.1 ± 1
KAR	68 ± 38	34 ± 34	46 ± 15	27 ± 13	0.6 ± 1
PES	71 ± 38	39 ± 34	52 ± 21	35 ± 19	1.5 ± 1
LAH	143 ± 69	71 ± 48	49 ± 25	42 ± 22	1.3 ± 1

Source: Dall'Osto 2012.
Note: $PM_{2.5}$ = particulate matter of less than 2.5 microns, SO_2 = sulfur dioxide, NO_2 = nitrogen dioxide, O_3 = ozone, CO = carbon monoxide.

Table 2.4 Maximum Values of $PM_{2.5}$, SO_2, NO_2, O_3, and CO for Five Different Cities: Peshawar (PES), Lahore (LAH), Karachi (KAR), Islamabad (ISL), and Quetta (QUE)

Maximum value	$PM_{2.5}$ (µg/m³)	SO_2 (µg/m³)	NO_2 (µg/m³)	O_3 (µg/m³)	CO (mg/m³)
ISL	157	32	196	148	5
QUE	96	136	83	72	4
KAR	201	173	122	86	2
PES	146	147	141	90	6
LAH	433	309	129	139	7

Source: Dall'Osto 2012.
Note: $PM_{2.5}$ = particulate matter of less than 2.5 microns, SO_2 = sulfur dioxide, NO_2 = nitrogen dioxide, O_3 = ozone, CO = carbon monoxide.

The analysis of the 2007–10 time series on sulfur dioxide (SO_2) confirmed that Lahore was the city with the highest SO_2 concentrations (74 ± 48 μg/m³), with maximum daily values of 309 μg/m³. Other cities presented very high values of SO_2: Quetta (54 ± 26 μg/m³), Karachi (34 ± 34 μg/m³), and Peshawar (39 ± 34 μg/m³). Overall, SO_2 values were increasing over the course of the study period (2007–10).

The annual nitrogen dioxide (NO_2) concentrations derived from the 48-hour data revealed that the current levels in the country are slightly higher than the WHO air quality guideline value of 40 μg/m³. The highest concentrations were in Peshawar (52 ± 21 μg/m³), Islamabad (49 ± 28 μg/m³), Lahore (49 ± 25 μg/m³), and Karachi (46 ± 15 μg/m³). Concentrations were somewhat lower in Quetta (37 ± 15 μg/m³). Results from an analysis of data from 2007 to 2011 show that concentrations of O_3 and CO were well within the WHO guidelines. See figure 2.3.

Source Apportionment of Air Quality Pollutants

A statistical analysis shows that the strongest correlations (expressed as R^2) among key parameters are found between $PM_{2.5}$ and CO, implying road traffic is a main source of fine PM in Pakistan (figure 2.4). Since $PM_{2.5}$ correlates better with CO than with SO_2 and NO_2, it is possible that fresh direct traffic emissions are important contributions to the fine particulate mass levels. Other sources,

Figure 2.3 Temporal Trend for $PM_{2.5}$, SO_2, NO_2, O_3, and CO for Five Pakistan Cities: Peshawar (PES), Lahore (LAH), Karachi (KAR), Islamabad (ISL), and Quetta (QUE)

a.

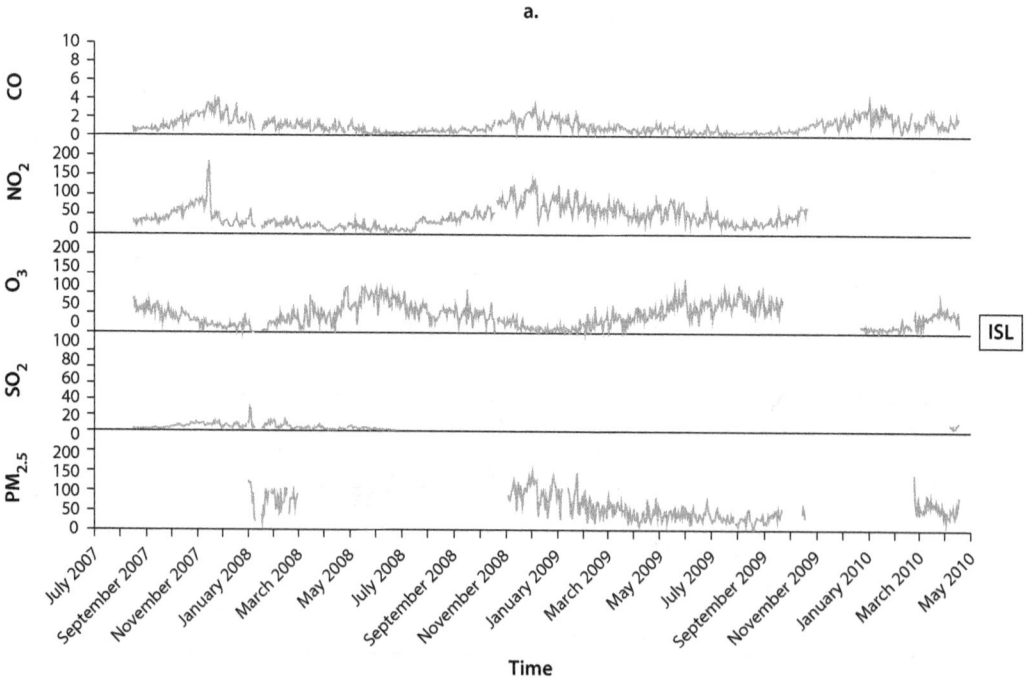

figure continues next page

Figure 2.3 Temporal Trend for PM$_{2.5}$, SO$_2$, NO$_2$, O$_3$, and CO for Five Pakistan Cities: Peshawar (PES), Lahore (LAH), Karachi (KAR), Islamabad (ISL), and Quetta (QUE) *(continued)*

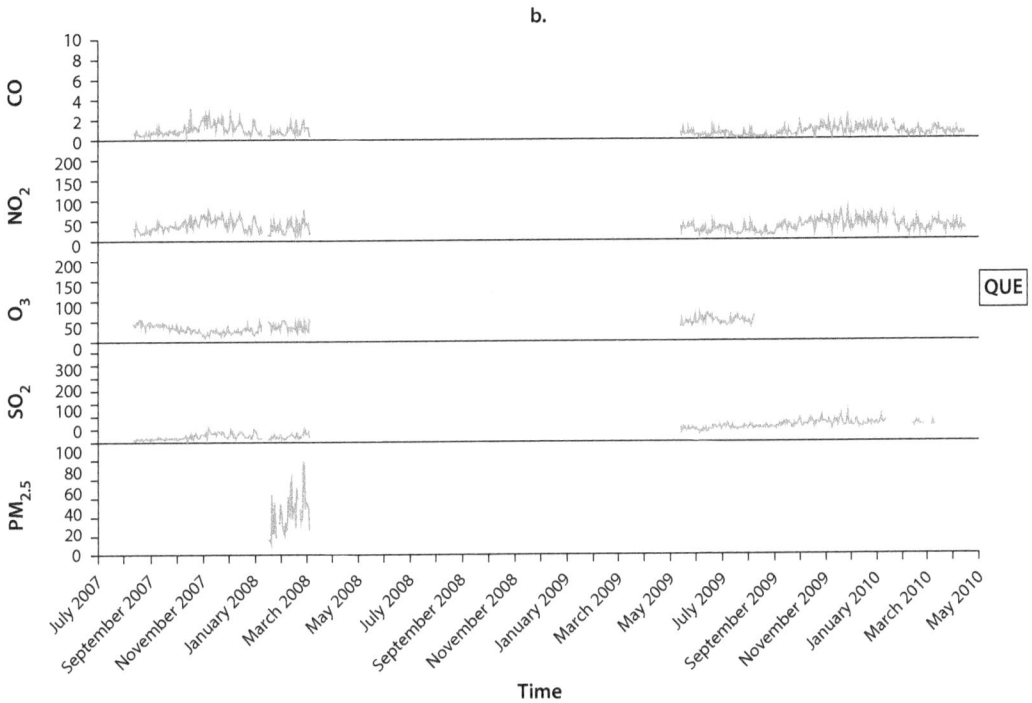

b.

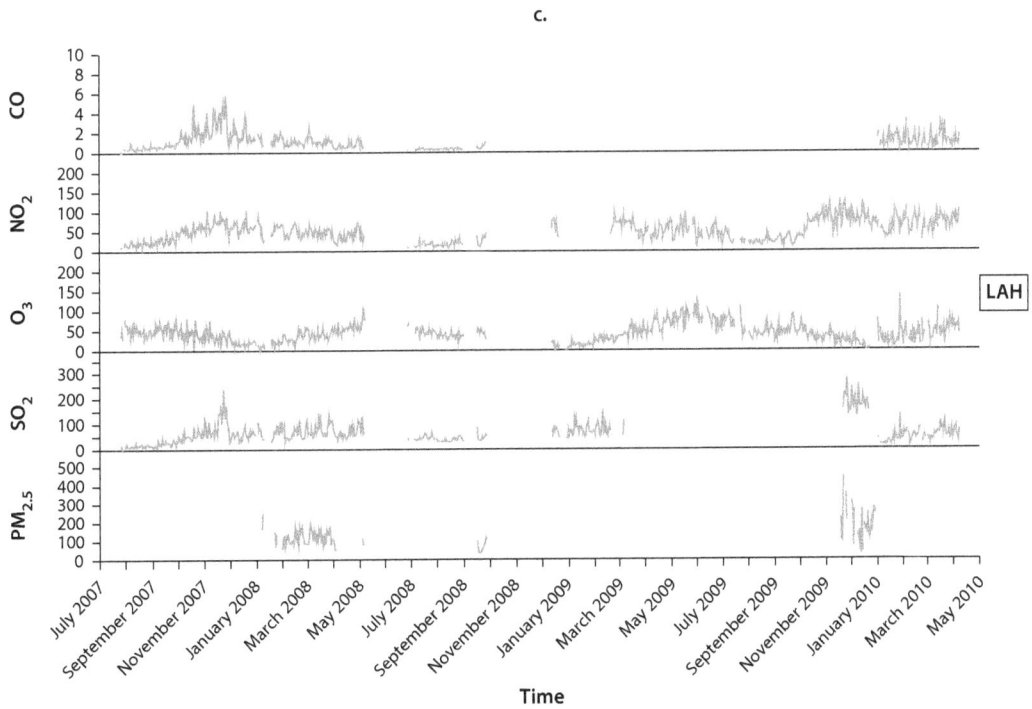

c.

figure continues next page

Figure 2.3 Temporal Trend for PM$_{2.5}$, SO$_2$, NO$_2$, O$_3$, and CO for Five Pakistan Cities: Peshawar (PES), Lahore (LAH), Karachi (KAR), Islamabad (ISL), and Quetta (QUE) *(continued)*

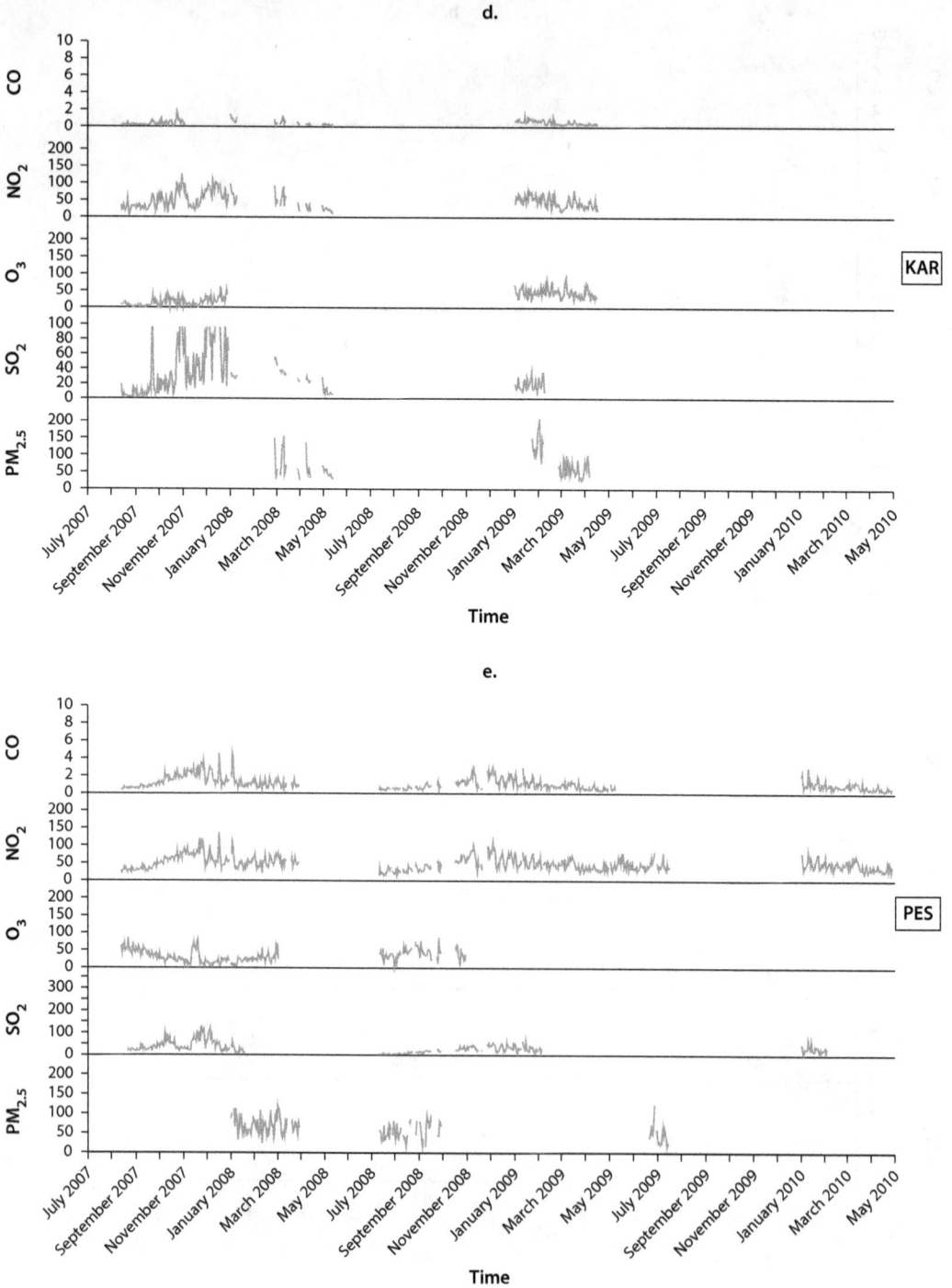

d.

e.

Source: Dall'Osto 2012.

Note: PM$_{2.5}$ = particulate matter of less than 2.5 microns, SO$_2$ = sulfur dioxide, NO$_2$ = nitrogen dioxide, O$_3$ = ozone, CO = carbon monoxide.

Figure 2.4 Correlations between PM$_{2.5}$, SO$_2$, NO$_2$, O$_3$, and CO for Islamabad (ISL), Quetta (QUE), Karachi (KAR), Peshawar (PES), and Lahore (LAH)

City	R^2	SO$_2$	NO$_2$	O$_3$	CO
ISL	PM$_{2.5}$	0.35	0.45	−0.37	0.7
	SO		0.5	−0.25	0.5
	NO$_2$			−0.2	0.4
	O$_3$				−0.4
QUE	PM$_{2.5}$	0.55	0.75	−0.4	0.85
	SO$_2$		0.7	−0.45	0.7
	NO$_2$			−0.55	0.85
	O$_3$				−0.5
KAR	PM$_{2.5}$	0.2	0.4	0	0.45
	SO$_2$		0.55	0	0.1
	NO$_2$			0	0.7
	O$_3$				0
PES	PM$_{2.5}$	0.2	0.5	0	0.6
	SO$_2$		0.4	0	0.4
	NO$_2$			0	0.8
	O$_3$				0
LAH	PM$_{2.5}$	0	0	0	0
	SO$_2$		0.6	0	0.6
	NO$_2$			0	0.6
	O$_3$				n.a.

Source: Dall'Osto 2012.
Note: n.a. = Not applicable, PM$_{2.5}$ = particulate matter of less than 2.5 microns, SO$_2$ = sulfur dioxide, NO$_2$ = nitrogen dioxide, O$_3$ = ozone, CO = carbon monoxide.

including industries and natural dust or sea salt may contribute too, but our analysis suggests that direct traffic emissions are more related to high concentrations of ambient aerosols. The low correlations obtained at Lahore may be due to the very high levels of PM (which may be associated with natural factors, such as dust and emissions from industrial and agricultural sources).

When considering factor analysis, the "Primary" factor includes traffic (NO$_2$, CO) and industrial (SO$_2$) gases, as well as PM$_{2.5}$. The "Primary" factor is mainly associated with primary anthropogenic emissions, and it represents the major component for the cities of Islamabad, Karachi, and Peshawar (33–43%). The "Secondary" factor is seen mainly during summer times, and it is associated with O$_3$ and high temperature values. It was the second main factor found (22–33%), and it is associated more with summer regional pollution events (table 2.5).

The analysis found that PM$_{2.5}$ is strongly correlated with CO and NO$_2$, indicating the importance of road traffic as a source, especially in winter months. By contrast, in the summer months, with higher wind speed, the influence of

Table 2.5 Principal Component Analysis: Results for Islamabad, Lahore, Karachi, and Peshawar

City	PCA factors		
	Primary	Secondary	Visibility
Islamabad			
$PM_{2.5}$	**0.9**	0.1	0.0
SO_2	**0.9**	0.2	0.0
NO_2	**0.9**	0.2	0.0
O_3	0.1	**0.8**	0.2
CO	**0.9**	0.0	0.1
Variance explained (%)	33	27	10
Karachi			
SO_2	**0.9**	0.1	0.3
NO_2	**0.9**	0.2	0.2
O_3	0.1	**0.9**	0.1
CO	**0.7**	0.5	0.1
Variance explained (%)	35	22	14
Peshawar			
$PM_{2.5}$	**0.7**	0.1	0.3
SO_2	**0.9**	0.3	0.2
NO_2	**0.9**	0.3	0.1
O_3	0.5	**0.6**	0.2
CO	**0.9**	0.0	0.1
Variance explained (%)	43	23	11
Lahore			
$PM_{2.5}$	**0.6**	0.3	0.3
SO_2	**0.8**	0.2	0.2
NO_2	**0.8**	0.5	0.2
O_3	0.2	**0.8**	0.1
CO	**0.8**	0.4	0.1
Variance explained (%)	22	33	11

Source: Dall'Osto 2012.
Note: $PM_{2.5}$ = particulate matter of less than 2.5 microns, SO_2 = sulfur dioxide, NO_2 = nitrogen dioxide, O_3 = ozone, CO = carbon monoxide. Numbers in bold are larger than the fit from the principal component analysis (PCA).

re-suspended surface dusts and soils and of secondary PM may play a bigger role. However, the strong correlation between road traffic markers (such as CO) and $PM_{2.5}$ suggests anthropogenic pollution is a major source of fine PM, leaving natural sources as minor ones.

Air Pollution and the Quality of Life in Pakistan

High concentrations of ambient air pollution are associated with lower gross domestic product (GDP) per capita levels and increased rates of mortality among young children. Generally speaking, countries with higher per capita income levels face lower levels of urban air pollution. However, Pakistan's PM_{10}

concentration levels are among the highest in the world, even above those of countries with lower levels of per capita income. Moreover, Pakistan's pollution levels are significantly higher than neighboring countries. For example, PM_{10} concentration levels in China and India are almost half of Pakistan's (figure 2.5).

Pakistan's largest cities remain among the least attractive in the world in terms of living conditions (including environmental quality) for expatriate executive managers (Mercer 2010). In 2010, Karachi, Lahore, and Islamabad ranked 191, 181, and 179, respectively, in 221 evaluated cities. Karachi, the only Pakistani city assessed in the Livability Index by the Economist Intelligence Unit, ranked 135 over a total of 140 cities (Economist Intelligence Unit 2010) (figure 2.6). Living conditions in selected Asian cities are perceived by private investors as correlated with foreign direct investments.[3]

Although specific indicators for environmental governance are not available, at least in a consistent and internationally comparable manner, the World Governance Indicators published on a yearly basis (World Bank 2010b) reflect some of the most rooted problems in Pakistan's institutional development, which are particularly relevant to the air quality management (AQM) framework. For instance, the *Government Effectiveness Index* captures perceptions of the quality of public services, the quality of the civil service, and the degree of its independence from political pressures, the quality of policy formulation and implementation, and the credibility of the government's commitment to such policies. Air quality positively correlated with governance. In 2008, Pakistan scored −0.80 on a scale from −2.5 to 2.5, among the 25% lowest performances in governance worldwide (World Bank 2010b).

Environmental quality indicators, including concentrations of PM, correlate with governance indicators, including government effectiveness, voice and

Figure 2.5 Income and Urban Pollution Levels across Countries, 2007

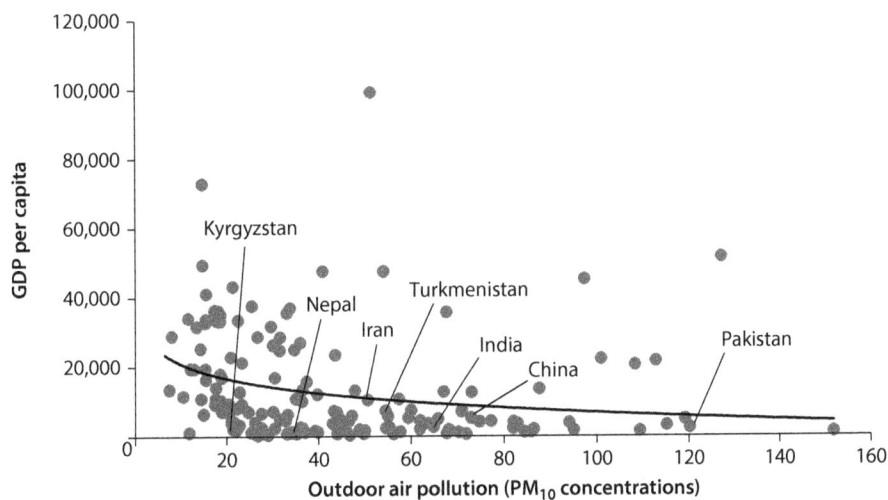

Source: World Bank 2010a.
Note: GDP per capita is measured in current US$. Outdoor air pollution refers to PM_{10} concentrations (micrograms per cubic meter) at the country level.

Figure 2.6 Quality of Living and Eco-Cities Indexes in Selected Cities, 2010

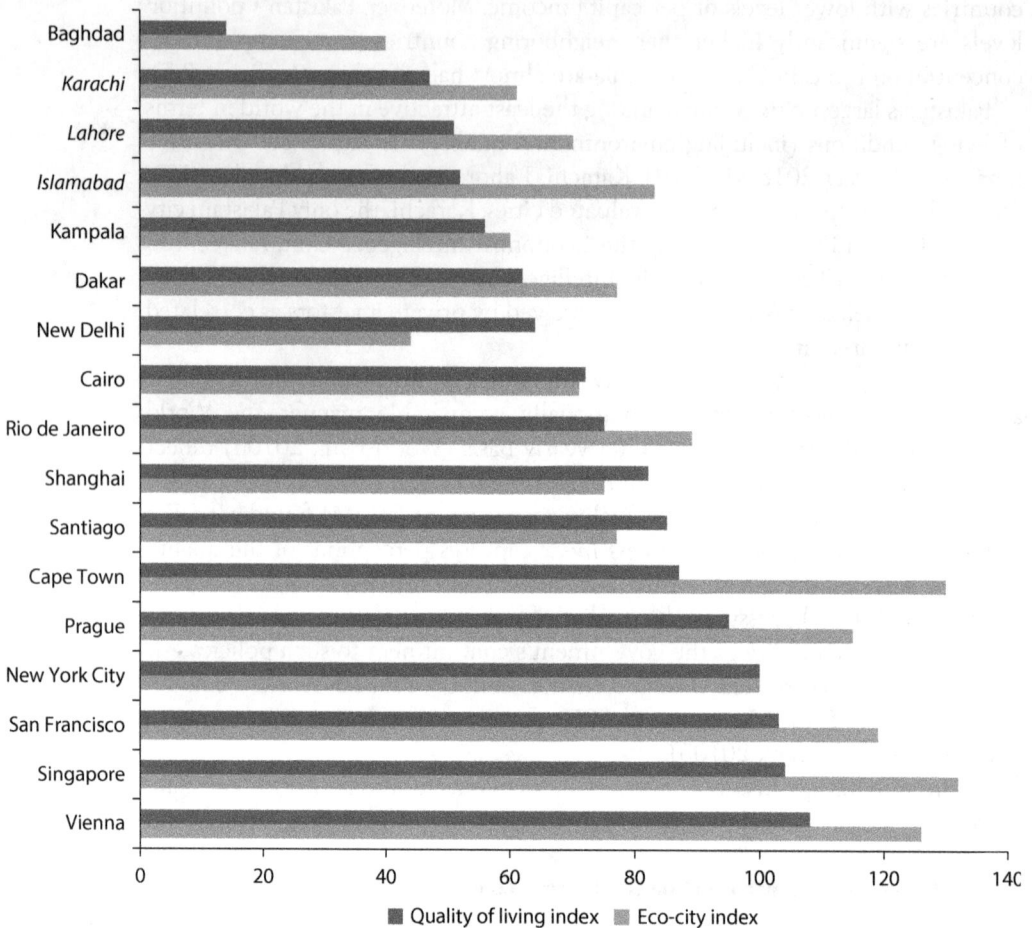

Sources: Economist Intelligence Unit 2010; Mercer 2010.
Note: Pakistani city names in italics.

accountability, political stability, regulatory quality, rule of law, and control of corruption (figure 2.7).

Limited available evidence indicates that concentrations of $PM_{2.5}$ in Islamabad, Karachi, Lahore, Peshawar, Rawalpindi, and Quetta are significantly above the National Environmental Quality Standards (NEQS) that came into force in July 2010, the stricter standards that came into force in January 2013, and the WHO guidelines (table 2.6).

As examples of international best practices, large metropolises, such as Mexico City, Santiago, and Bangkok have successfully reduced their ambient concentrations of $PM_{2.5}$ to a level that is even lower than those of small Pakistani cities such as Gujranwala (Lodhi 2009).

The sources of air pollution in Pakistan include mobile sources, such as heavy-duty vehicles, and stationary sources, such as power plants, burning of waste, and

Figure 2.7 Government Effectiveness and Air Quality (PM$_{10}$), 2008

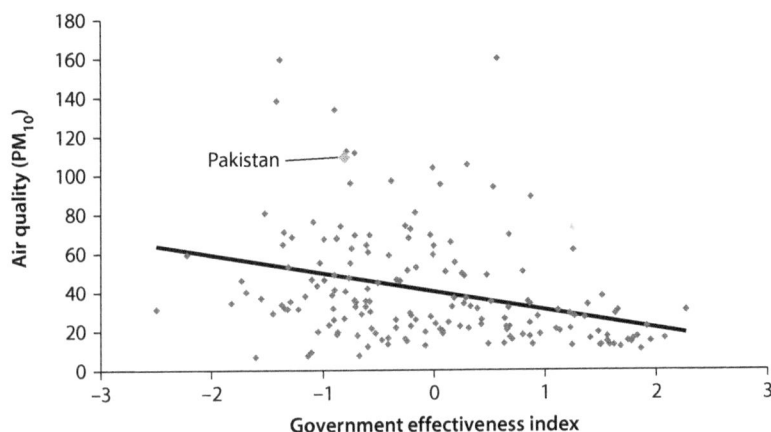

Sources: World Bank 2010a, 2010b.
Note: PM$_{10}$ = particulate matter of less than 10 microns.

Table 2.6 Concentration Levels (Daily Average) of Suspended PM$_{2.5}$ in Pakistan
μg/m³

Authors	Area	Year	Islamabad	Lahore	Karachi	Peshawar	Quetta
Pak-EPA 2007	6 sites	2005	—	—	—	—	104–222
Husain, Farhana, and Ghauri 2007	—	2005–06	—	53–476	—	—	—
Ghauri 2010[a]	Mobile stations	2007	43.7	—	—	—	—
	Fixed stations	2007	47.2	74.6	71.7	185.5	206.4

Sources: Colbeck, Zaheer, and Zulfiqar 2010; Ghauri 2010 from data collected at stations installed by the Japanese International Cooperation Agency.
Note: — = not measured. PM$_{2.5}$ (particulate matter of less than 2.5 microns) World Health Organization Guidelines = 10 μg/m³; European Union Ambient Air Quality Standards = 25 μg/m³; United States Ambient Air Quality Standards = 25 μg/m³; Pakistan 2010 National Environmental Quality Standards = 25 μg/m³.
a. The results of Ghauri (2010) are part of a World Bank-financed study reporting data obtained through interviews and primary data collected from monitoring stations in Pakistan.

natural dust. The transport and energy sectors contribute more than two-thirds of PM, nearly half of NO$_x$, two-thirds of CO, and about half of hydrocarbon emissions. PM is the pollutant associated with the largest economic damage due to its effect on human health in Pakistan (Colbeck, Zaheer, and Zulfiqar 2010; Ghauri 2010; Ilyas 2007).[4] See figure 2.8.

In Pakistan, as in many other countries, industries tend to concentrate in a few urban centers where their competitiveness is enhanced because of the availability of specialized labor, inter-industry spillovers, higher road density, local transfer of knowledge, and access to international supplier and buyer networks. These factors largely explain the clustering of large-scale manufacturing and high associated employment levels around the metropolitan areas of Karachi and Lahore. The process of industrialization will induce unprecedented urban growth over the coming decades. Industrial growth and urbanization in Pakistan will aggravate "public bads" (such as air pollution or traffic congestion).

Figure 2.8 Measured Concentration Levels of PM$_{10}$ in Selected Cities in Pakistan; Data from Studies Undertaken between 1999 and 2006

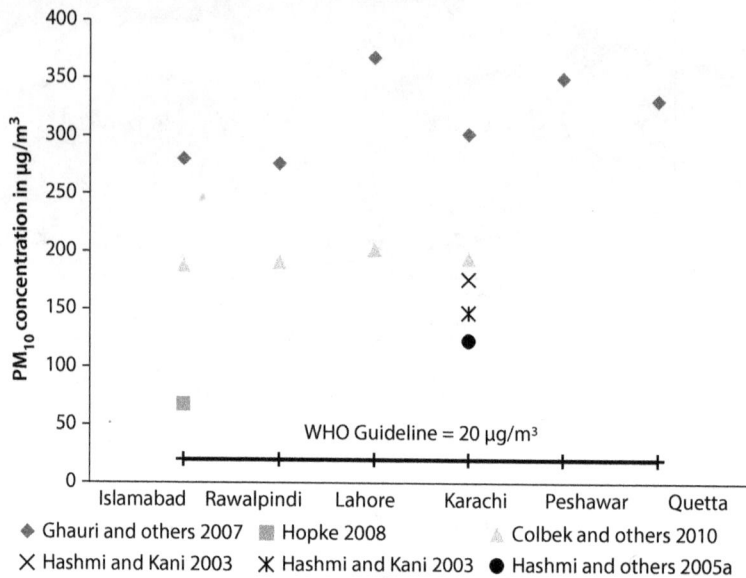

Sources: Colbeck, Zaheer, and Zulfiqar 2008, Ghauri 2008, 2010; Ghauri, Lodhi, and Mansha 2007; Hashmi and Khani 2003; Hashmi and others 2005a, 2005b; Hopke and others 2008; Husain, Farhana, and Ghauri 2007; Pak-EPA 2007; SUPARCO 2004.

Note: PM$_{2.5}$ = particulate matter of less than 2.5 microns; PM$_{10}$ = particulate matter of less than 10 microns; WHO = World Health Organization.

Based on available data, it is estimated that mobile sources are the largest contributor of air pollutants, particularly PM$_{2.5}$, in large urban centers in Pakistan.[5] Preliminary estimates indicate that the road transport sector is responsible for 85% of PM$_{2.5}$ total emissions and 72% of the PM$_{10}$ emissions. While other sectors, particularly industry, also contribute to Pakistan's severe urban air pollution, preliminary estimates indicate that the vast majority of PM emissions stem from the road transportation sector (figure 2.9). Given that PM, and particularly PM$_{2.5}$, has been linked with negative health effects, these figures provide the prima facie evidence for tackling vehicle sources as one of the top priorities in Pakistan's AQM efforts.

The Economic Cost of Air Pollution in Pakistan

Ambient air quality problems tend to be most severe in urban areas, where population, pollution sources, automobiles, and industry are most concentrated. In Pakistan, more than 35% of the population lives in urban areas, most of them in cities of more than 1 million inhabitants. Air pollution is associated with increased respiratory illness and premature mortality. By 2005, the direct cost of the damage associated with outdoor air pollution was estimated at 1.1% of Pakistan's GDP or US$1.07 billion.[6,7] These costs include estimates from premature mortality and morbidity associated with cardiovascular and respiratory diseases, lower

Figure 2.9 Sources of Particulate Matter Emissions from Fossil Fuels Combustion in Pakistan, 2008

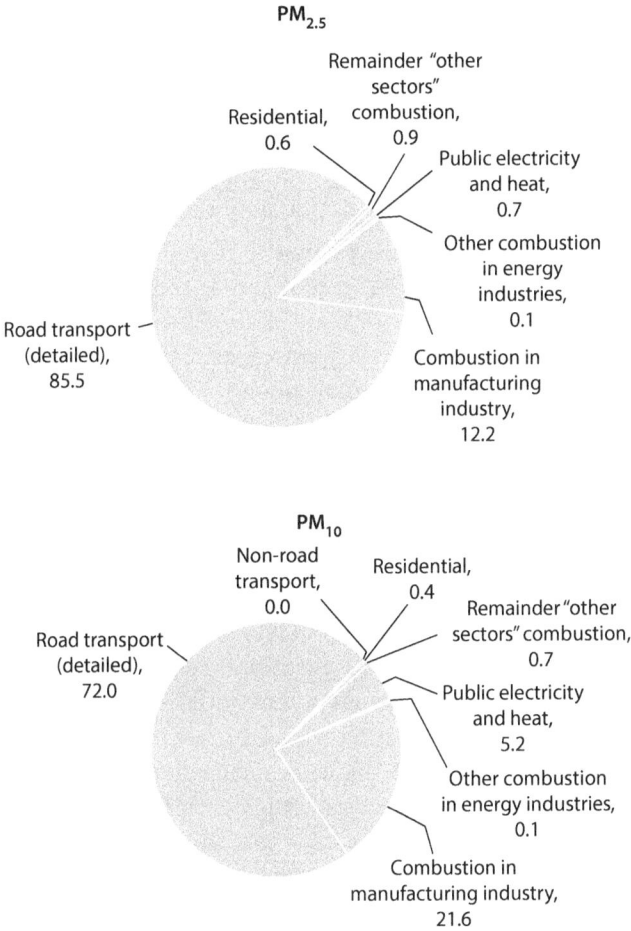

Percent

PM$_{2.5}$

- Remainder "other sectors" combustion, 0.9
- Residential, 0.6
- Public electricity and heat, 0.7
- Other combustion in energy industries, 0.1
- Road transport (detailed), 85.5
- Combustion in manufacturing industry, 12.2

PM$_{10}$

- Non-road transport, 0.0
- Residential, 0.4
- Remainder "other sectors" combustion, 0.7
- Road transport (detailed), 72.0
- Public electricity and heat, 5.2
- Other combustion in energy industries, 0.1
- Combustion in manufacturing industry, 21.6

Source: Estimated using the emissions inventory methodology of the Malé Declaration on Control and Prevention of Air Pollution http://www.rrcap.unep.org/male.

Note: PM$_{2.5}$ = particulate matter of less than 2.5 microns; PM$_{10}$ = particulate matter of less than 10 microns.

respiratory illness (LRI) in children, and other minor related costs.[8–10] However, when indirect costs, such as the well-known effects of air pollution-related illnesses on education, malnutrition, and earnings, are taken into account, the damage cost of ambient air pollution could be even higher (World Bank 2008a).

At that time, more than 22,600 deaths per year were directly or indirectly attributable to ambient air pollution in Pakistan, of which more than 800 are children under five years of age (World Bank 2006, 2008a). Outdoor air pollution alone caused more than 80,000 hospital admissions per year, nearly 8,000 cases of chronic bronchitis, and almost 5 million cases of lower respiratory cases among children under five. To put these numbers in perspective, the harm done by air pollution exceeds most other high-profile causes of mortality and morbidity that receive significantly more attention in Pakistan, including road accidents, which resulted in over 5,500 reported fatalities and nearly 13,000 non-fatal injuries in 2007 (WHO 2011).[11]

In Punjab alone, outdoor air pollution had a mean annual cost of PRs 9–35 billion, or 0.24–0.95% of the province's GDP in 2007. The low bound reflects the use of human capital approach for valuation of mortality, and the high bound reflects the use of value of statistical life. About 83% of the mean cost is associated with mortality, and 17% with morbidity. Urban air pollution results in 137,380 Disability-Adjusted Life Years (DALYs),[12] out of which mortality represents 49% and morbidity 51% (Strukova 2008).

These $PM_{2.5}$ concentrations are estimated to cause over 9,000 premature deaths each year, representing 20% of acute lower respiratory infection (ALRI) mortality among children under five years of age, 24% of cardiopulmonary mortality, and 41% of lung cancer mortality among adults 30 or more years of age in these cities. About 12% of the deaths are among children under five years of age and 88% are among adults. Nearly 80% of the deaths are in Karachi.[13] PM concentrations are also the estimated cause of 59% of chronic bronchitis cases in these cities, or a total of nearly 185,000 cases, and nearly 33,000 hospital admissions, over 645,000 emergency room visits, over 1.6 million cases of ALRI in children, over 100 million restricted activity days, and over 300 million respiratory symptoms annually.[14] These annual health effects represent 203,000 DALYs, of which 97,000 are from premature mortality and 106,000 from morbidity.

O_3 pollution affects the economic yield of major agricultural crops. Several studies show that O_3 pollution causes a substantial effect on agricultural production in Pakistan. By comparing crops grown in three different plots (charcoal-filtered air, unfiltered air, and unchambered field plots), Wahid (2006) reported seed yield reductions ranging from 18% to 43% for three different wheat varieties. By using a similar methodology, Ahmed (2009) found losses that were even more significant involving two commonly grown mung bean varieties (47–51%). Limited knowledge on the actual levels of exposure of crops to the effects of O_3 makes it difficult to estimate the overall economic losses attributable to air pollution. Because O_3 effects do not remain localized, but instead travel long distances in the atmosphere, the potential adverse effects are likely to be found not only in peri-urban areas crops, but also in extensively cultivated rural areas.[15] However, Van Dingenen and others (2008) estimate that the economic cost from O_3 damage to crops in Pakistan might amount to US$550 million.

The Economic Cost of Karachi's Air Pollution

Karachi ranks 135 out of 140 in the world's urban livability index (based on stability, health care, culture, and environment, education, and infrastructure) from the Economist Intelligence Unit (2010). The high levels of dangerous pollutants, such as fine and ultrafine PM emitted from highly polluting vehicles (particularly trucks), cause significant health risks to urban populations. PM released into the air, particularly PM having a diameter of less than

2.5 microns (PM$_{2.5}$), is one of the key causes of poor health outcomes in Pakistan.[16]

The population of Sindh province was around 35.7 million in 2009. An estimated 17 million people lived in 13 cities with a population greater than 100,000 inhabitants, of which 13 million were in Karachi (World Gazetteer 2011). Much of this population is exposed to outdoor ambient air concentrations of PM that are many times higher than recent WHO guidelines. Annual average PM$_{2.5}$ ambient air concentrations in Karachi are estimated at 88 micrograms per cubic meter (µg/m^3) (figure 2.10), and in the range of 55–88 µg/m^3 in other cities with a population greater than 100,000 inhabitants. The analysis focuses on PM and especially PM$_{2.5}$ because it is the outdoor air pollution globally associated with the largest health effects (WHO 2004). Fine PM is well documented to have a robust association with several serious public health effects (for example, significant increase in cardiovascular and pulmonary diseases that may result in death or permanent disability). As mentioned above, nearly 80% of the more than 9,000 premature deaths caused every year in Pakistan by high PM$_{2.5}$ concentrations are in Karachi.

Although trucks represent a minor fraction of Pakistan's vehicle fleet, they are a major source of pollutants of local concern. As the number of registered vehicles increases in Pakistan, so does the level of air pollution in urban areas, particularly in densely populated metropolitan regions such as Karachi. They emit a large share of various pollutants, including PM, sulfur dioxide (SO$_2$), and nitrous oxides (NO$_x$), which contribute to respiratory ailments.

Sources of lead (Pb) exposure in Sindh today include industry and workshops, dust and soil, food and fish, drinking water, housing materials, paint, cosmetics (for example, surma[17]), utensils, children's toys, and other materials and articles

Figure 2.10 Estimate of Annual Average PM$_{2.5}$ Ambient Air Concentrations in Karachi, 2006–09 (µg/m^3)

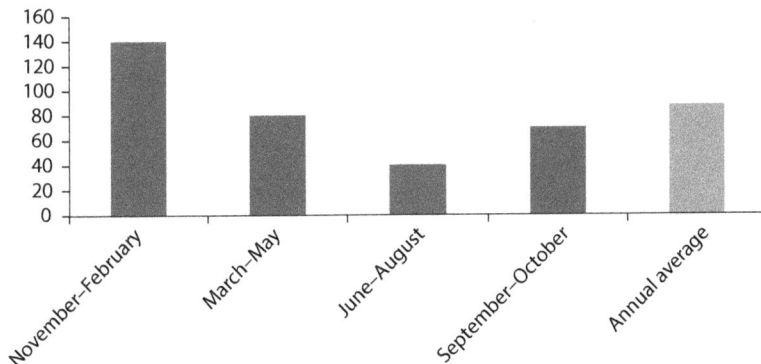

Source: Sánchez-Triana and others 2014, based on Alam and others 2010, Ghauri 2008, Mansha and others 2011, Sindh EPA 2010.
Note: PM$_{2.5}$ = particulate matter of less than 2.5 microns.

Cleaning Pakistan's Air • http://dx.doi.org/10.1596/978-1-4648-0235-5

containing or being contaminated by Pb. Young children are the population group that is most vulnerable to lead exposure. Pb levels in children's blood in Karachi have declined substantially in the last two decades with the abolishment of Pb in gasoline, possibly from over 35 micrograms of Pb per deciliter of blood (µg/dL) in the late 1980s to around 15 µg/dL in 2000 (Kadir and others 2008). Recent data on children's blood lead levels (BLL) in Sindh and Pakistan are scarce. Based on data in Khan, Ansari, and Khan (2011) and Ahmad and others (2009), it may be suspected that mean BLL in children under five years of age is 7 µg/L in urban areas and 5 µg/dL in rural areas of Sindh, that is, half of BLLs in 2000 (figure 2.11).

In a study of drinking water in 18 districts of Karachi in 2007/08, Pb concentrations in 89% of groundwater sources and tap water from surface water sources exceeded the WHO guideline limit of 10 micrograms per liter of water (µg/L). The mean lead concentration was 77 µg/L in drinking water originating from surface water sources and 146 µg/L in groundwater (Ul-Haq and others 2011). In a study of groundwater quality in locations throughout Sindh province other than Karachi, 54% of samples (mainly hand pumps and dug wells) contained lead concentration exceeding the WHO guideline of 10 µg/L, and 23% of the samples contained lead in concentrations greater than 50 µg/L. The highest measured concentration was 111 µg/L (Junejo n.d.).[18] Based on these Pb concentrations, it is estimated in this book that Pb in drinking water results in an average BLL of 3–4 µg/dL among children <5 years old in Sindh. Thus, this one source of lead exposure may be responsible for over 50% of suspected BLLs among these children.

Several recent international studies have documented neuropsychological effects in children, such as impaired intelligence measured as intelligence quotient (IQ) losses even at BLLs well below 10 µg/dL (Canfield and others 2003; Jusko and others 2008; Lanphear and others 2005; Surkan and others 2007).

Figure 2.11 Suspected Blood Lead Levels (BLL) in Children under Five Years in Sindh

Percentage of children

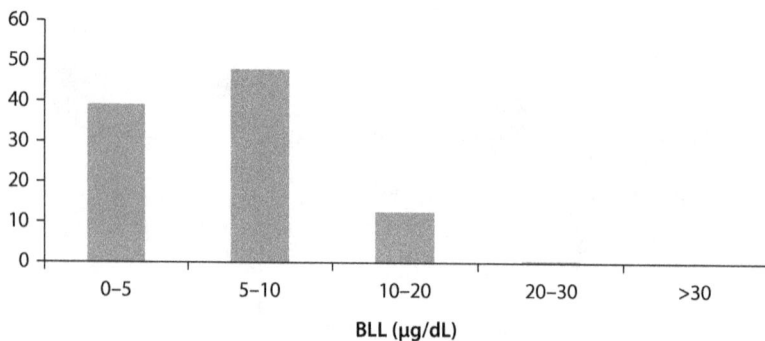

BLL (µg/dL)

Source: Sánchez-Triana and others 2014.

Figure 2.12 Loss of IQ Points in Early Childhood in Relation to Lower Threshold Levels (X_0) of Blood Lead Levels (BLL)

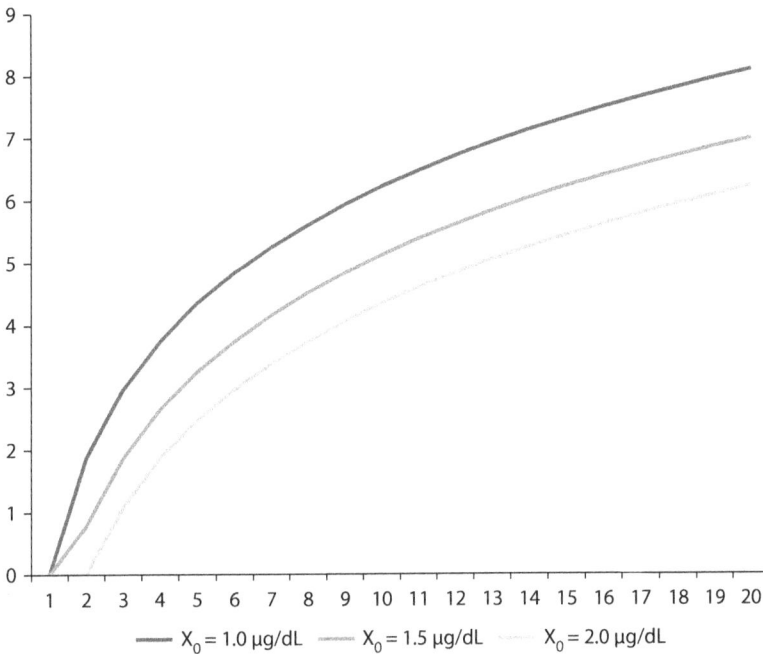

X₀ = 1.0 μg/dL ── X₀ = 1.5 μg/dL ── X₀ = 2.0 μg/dL

Source: Sánchez-Triana and others 2014.
Note: IQ = intelligence quotient.

Moreover, these studies have not identified a threshold below which BLLs have no impact on children's intelligence. Thus, according to this recent evidence, a child 5–7 years of age with a BLL of 10 μg/dL has lost as many as 4.4–6.2 IQ points due to Pb exposure during the first five years of life if the lower BLL threshold of effects is in the range of 1–2 μg/dL (figure 2.12). The annual loss of IQ points in children in Sindh is consequently estimated to be 2 million if the threshold is 2 μg/dL, and as much as 3.7 million if the threshold is 1 μg/dL.[19]

Conclusions

Air pollution in Pakistan's cities is the worst in the South Asia Region. The concentrations of PM in Pakistan's cities are much higher than those experienced in the urban areas of countries such as Bhutan, India, and Sri Lanka. In addition, while these countries have adopted a number of measures that have resulted in reduced urban air pollution, Pakistan has yet to follow suit and is paying the costs of increasingly high outdoor air pollution.

Quality of life in Pakistan's cities is low, and these urban centers are among the least attractive for expatriate executive managers. Some surveys suggest that this situation can increase difficulties for firms to recruit and retain professionals, as well as to invest directly in those cities. This is one of the many issues illustrating the linkages between environmental quality and competitiveness.

PM is the most damaging air pollutant to human health in Pakistan. Concentrations of air pollutants in Pakistan's large cities are significantly above the NEQS that came into force in July 2010, the stricter standards that came into force in 2013, and the guidelines recommended by WHO.

Notes

1. PM of less than 10 microns in diameter (PM_{10}) is capable of penetrating deep into the respiratory tract and causing damage. The World Bank (2010a) Little Green Data Book estimates PM as an indicator for a particular country as the population weighted average of all cities in a particular country with a population greater than 100,000. Fine PM ($PM_{2.5}$) has become increasingly important because of its stronger association with negative health effects. However, the graph shows PM_{10} values because these have been monitored in a wider range of countries and for longer periods of time than $PM_{2.5}$, thereby allowing for better comparisons across geographical areas and time scales.

2. According to Pope and Dockery (2006, 710–11), PM air pollution is an air-suspended mixture of solid and liquid particles that include coarse particles, fine particles, and ultrafine particles. Coarse particles include dust, soil, and other crustal materials from roads, farming, mining, windstorms with "mass concentrations of particles greater than a 2.5 μm cut-point." Fine particles are derived primarily from direct emissions from combustion processes, such as vehicle use of gasoline and diesel, wood burning, coal burning for power generation, and industrial processes. Fine particles also consist of transformation products, including sulfate and nitrate particles, which are generated by conversion from primary sulfur and nitrogen oxide emissions and secondary organic aerosol from VOC emissions. The most common indicator of fine PM is $PM_{2.5}$, consisting of particles with an aerodynamic diameter less than or equal to a 2.5 μm cut point. Ultra-fine particles are typically defined as particles with an aerodynamic diameter less than 0.1 μm. Ambient air in urban and industrial environments is constantly receiving fresh emissions of ultrafine particles from combustion-related sources, such as vehicle exhaust and atmospheric photochemical reactions. "With regard to $PM_{2.5}$, various toxicological and physiological considerations suggest that fine particles may play the largest role in affecting human health. For example, they may be more toxic because they include sulfates, nitrates, acids, metals, and particles with various chemicals adsorbed onto their surfaces. Furthermore, relative to larger particles, particles indicated by $PM_{2.5}$ can be breathed more deeply into the lungs, remain suspended for longer periods of time, penetrate more readily into indoor environments, and are transported over much longer distances."

3. A 2007 survey by the American Chamber of Commerce in Hong Kong SAR, China, found that 57% of surveyed companies indicated that the current trend and changes in Hong Kong SAR, China's air pollution might cause their companies to invest elsewhere. In addition, 60% of the interviewed entrepreneurs stated that better air quality conditions would lead to increasing investments in the city by their own company. The survey also shows investors' concerns about the difficulties in recruiting and retaining professionals in Hong Kong SAR, China, due to poor environmental conditions (American Chamber of Commerce 2007, 4).

4. Ghauri's (2010) results are part of a World Bank-financed study that report data obtained through interviews and primary data collected from monitoring stations in Pakistan.

5. Comprehensive emissions inventories covering a wide range of stationary sources and other mob SAR, China, ile sources are needed for reliable fine and ultrafine PM estimates and source apportionment.

6. Cost of mortality is based on the human capital approach for children and the value of statistical life for adults. Due to the lack of willingness-to-pay studies in Pakistan, the cost-of-illness approach (mainly medical cost and value of time losses) was applied as the only measure of morbidity costs.

7. High concentrations of fine particulate matter $PM_{2.5}$ (one of the main pollutants caused by fossil fuels combustion) are associated with the development of pneumonia, which is one of the main causes of infant mortality in Pakistan. Exposure to air pollutants at early ages increases the likelihood of contracting pneumonia (Grigg 2009), whereas aggregate data show a positive (although somewhat weak) cross-country correlation between PM_{10} concentration levels and the incidence of pneumonia as a cause of death among children under five.

8. Jamison, Lau, and Wang (1998), analyzing data from 53 countries from 1965 to 1990, estimated the impact on income of a number of health-related variables, and concluded that around 8% of total growth in per capita income during that time was due to improvements in adult survival rates. Bloom and Williamson (1998) conclude that one-third to one-half of the growth experienced from 1965 to 1990 in East Asia can be attributed to reduced child mortality and the consequent reduction in fertility.

9. While these effects are mainly focused on adult health, they can be significantly reinforced by the demographic changes resulting from reduced child mortality.

10. Countries with the weakest health conditions have a much harder time achieving sustained growth than do countries with better health conditions (Belli and Appaix 2003).

11. Road safety data can be found at http://www.who.int/violence_injury _prevention/road_safety_status/ country_profiles/pakistan.pdf.

12. DALYs are a standard measure that combines disparate health effects using a consistent common denominator. DALYs adjust the years of healthy life lost to illness and premature mortality, with a weighting function that corrects for the impacts of death and illness at different ages.

13. Mortality is estimated based on Ostro (2004) and on Pope and others (2002, 2009).

14. Morbidity is estimated based on Ostro (1994) and on Abbey and others (1995).

15. In the United States, annual economic losses of major agricultural crops due to yield depression by air pollution from motor vehicles would amount to $3–6 billion annually (Murphy and others 1999, 287).

16. PM is the term for airborne particles, including dust, dirt, soot, smoke, and liquid droplets. Particles can be suspended in the air for long periods. Some particles are directly emitted into the air. They come from a variety of sources such as vehicle exhaust, factories, construction sites, tilled fields, unpaved roads, stone crushing, and burning of wood. Particles also can be created by atmospheric conversion of SO_2 and NO_x into sulfates and nitrates. Most measurements of PM in Pakistan are of total suspended particles (TSP). There is strong scientific evidence that elevated concentrations of fine and ultrafine PM of less than 2.5 microns ($PM_{2.5}$) and 1.0 microns ($PM_{1.0}$) pose an even greater health risk than PM of less than 10 microns (PM_{10}). However, no systematic monitoring information on $PM_{2.5}$ or $PM_{1.0}$ is available at this time in Pakistan.

17. http://www.fda.gov/Cosmetics/ProductandIngredientSafety/ProductInformation/ucm137250.htm.

18. The study is undated but published in 2005 or more recently. The study does not state the year of the groundwater sampling.

19. Based on the log-linear central estimate of effects of BLLs in the meta-analysis by Lanphear and others (2005).

Bibliography

Abbey, D., M. Lebowitz, P. Mills, and F. Petersen. 1995. "Long-Term Ambient Concentrations of Particulates and Oxidants and Development of Chronic Disease in a Cohort of Nonsmoking California Residents." *Inhalation Toxicology* 7: 19–34.

Ahmad, T., A. Mumtaz, D. Ahmad, and N. Rashid. 2009. "Lead Exposure in Children Living around the Automobile and Battery Repair Workshops." *Biomedica*, 25: 128–32.

Ahmed, S. 2009. "Effects of Air Pollution on Yield of Mungbean in Lahore, Pakistan." *Pakistan Journal of Botany* 41 (3): 1013–21. http://www.pakbs.org/pjbot/PDFs/41%283%29/PJB41%283%291013.pdf.

Alam, K., M. J. Iqbal, T. Blaschke, S. Qureshi, and G. Khan. 2010. "Monitoring Spatio-temporal Variations in Aerosols and Aerosol-cloud Interactions over Pakistan Using MODIS Data." *Advances in Space Research* 46, 1162–76.

American Chamber of Commerce. 2007. "2007 Environment Survey." http://www.amcham.org.hk/images/pressCenter/Reports/2007_environment_survey.doc.

Belli, P. C., and O. Appaix. 2003. "The Economic Benefits of Investing in Child Health." Health, Nutrition, and Population World Bank Discussion Papers, 29256, World Bank, Washington, DC. http://www-wds.worldbank.org/external/default/WDSContentServer/WDSP/IB/2004/06/08/000265513_20040608124402/Rendered/PDF/292560Economic0Benefits0of0Investing.pdf.

Bloom, D. E., and J. G. Williamson. 1998. "Demographic Transitions and Economic Miracles in Emerging Asia." *World Bank Economic Review* 12 (3): 419–55. http://www.nber.org/papers/w6268.pdf?new_window=1.

Canfield, R. L., C. R. Henderson, D. A. Cory-Slechta, C. Cox, T. A. Jusko, and B. P. Lanphear. 2003. "Intellectual Impairment in Children with Blood Lead Concentrations below 10 μg per Deciliter." *New England Journal of Medicine* 348 (16): 1517–26.

Colbeck, I., A. Zaheer, and A. Zulfiqar. 2010. "The State of Ambient Air Quality in Pakistan: A Review." *Environmental Science and Pollution Research* 17: 49–63. http://www.springerlink.com/content/f718jn535422j0wh/fulltext.pdf.

Dall'Osto, M. 2012. Unpublished consultant report for the World Bank, Washington, DC.

Economist Intelligence Unit. 2010. "Livability and Ranking Overview. Worldwide Cost of Living Survey." http://store.eiu.com/product/475217632.html.

Ghauri, B. 2008. "Satellite Data Applications in Atmospheric Monitoring." SUPARCO, presented at the United Nations/Austria/European Space Agency Symposium, Graz, Austria, September 9–12.

———. 2010. *Institutional Analysis of Air Quality Management in Urban Pakistan.* Study commissioned by the World Bank. Washington, DC: World Bank. cleanairinitiative.org/portal/system/files/.../AQM_Draft_Final_Report.pdf.

Ghauri, B., A. Lodhi, and M. Mansha. 2007. "Development of Baseline (Air Quality) Data in Pakistan." *Environmental Monitoring and Assessment* 127: 237–52.

Grigg, J. 2009. "Particulate Matter Exposure in Children. Relevance to Chronic Obstructive Pulmonary Disease." *The Proceedings of the American Thoracic Society* 6: 564–69. http://pats.atsjournals.org/cgi/content/short/6/7/564.

Hashmi, D. R., and M. I. Khani. 2003. "Measurement of Traditional Air Pollutants in Industrial Areas of Karachi, Pakistan." *Journal of the Chemical Society of Pakistan* 25: 103–09. http://jcsp.org.pk/index.php/jcsp/article/viewFile/1413/966.

Hashmi, D. R., G. H. Shaikh, and T. H. Usmani. 2005a. "Ambient Air Quality at Port Qasim in Karachi City." *Journal of the Chemical Society of Pakistan* 27: 575–79.

———. 2005b. "Air Quality in the Atmosphere of Karachi City—An Overview." *Journal of the Chemical Society of Pakistan* 27: 6–13. http://jcsp.org.pk/index.php/jcsp/article /viewFile/820/465.

Hopke, P. K., D. D. Cohen, B. A. Begum, S. K. Biswas, B. Ni, G. G. Pandit, M. Santoso, Y. Chung, P. Davy, A. Markwitz, S. Waheed, N. Siddique, F. L. Santos, P. C. B. Pabroa, M. C. S. Seneviratne, W. Wimolwattanapun, S. Bunprapob, T. B. Vuong, P. D. Hein, and A. Markowicz. 2008. "Urban Air Quality in the Asian Region." *Science of the Total Environment* 404: 103–12. http://www.sciencedirect.com/science?_ob=MImg& _imagekey=B6V78-4T3CPYB-1-S&_cdi=5836&_user=145269&_pii =S0048969708005858&_origin=&_coverDate=10%2F01%2F2008&_sk=995959998 &view=c&wchp=dGLbVzz-zSkWW&md5=111219ef0c510b007ffcae4c7a8a71b0 &ie=/sdarticle.pdf.

Husain, L., B. K. Farhana, and B. M. Ghauri. 2007. "Emission Sources and Chemical Composition of the Atmosphere of a Mega-City in South Asia." *Eos Trans.* AGU 88 (23), Joint Assembly Supplement, A31C–01. http://adsabs.harvard.edu/abs /2007AGUSM.A31C..01H.

Ilyas, S. Z. 2007. "A Review of Transport and Urban Air Pollution in Pakistan." *Journal of Applied Sciences and Environmental Management* 11 (2): 113–21. http://www.ajol .info/index.php/jasem/article/viewFile/55004/43484.

Jamison, D. T., L. J. Lau, and J. Wang. 1998. "Health's Contribution to Economic Growth, 1965–1990." In *Health, Health Policy and Economic Outcomes, Final Report of the Health and Development Satellite.* Geneva, Switzerland: World Health Organization.

Junejo, S. A. n.d. "Groundwater Quality in Sindh." Indus Institute for Research and Education (IIRE), Hyderabad, Pakistan.

Jusko, T. A., C. R. Henderson, Jr., B. P. Lanphear, D. A. Cory-Slechta, P. J. Parsons, and R. L. Canfield. 2008. "Blood Lead Concentrations < 10 ug/dL and Child Intelligence at 6 Years of Age." *Environmental Health Perspectives* 116 (2): 243–48.

Kadir, M. M., N. Z. Janjua, S. Kristensen, Z. Fatmi, and N. Sathiakumar. 2008. "Status of Children's Blood Lead Levels in Pakistan: Implications for Research and Policy." *Public Health* 122 (7): 708–15.

Khan, D. A., W. M. Ansari, and F. A. Khan. 2011. "Synergistic Effects of Iron Deficiency and Lead Exposure on Blood Lead Levels in Children." *World Journal of Pediatrics* 7 (2): 150–54.

Lanphear, B. P., R. Hornung, J. Khoury, K. Yolton, P. Baghurst, D. C. Bellinger, R. L. Canfield, K. N. Dietrich, R. Bornschein, T. Greene, S. J. Rothenberg, H. L. Needleman, L. Schnaas, G. Wasserman, J. Graziano, and R. Roberts. 2005. "Low-level Environmental Lead Exposure and Children's Intellectual Functions: An International Pooled Analysis." *Environmental Health Perspectives* 113 (7): 894–99.

Larsen, B., and J. M. Skjelvik. 2012. *Environmental Health Priorities in the Province of Sindh, Pakistan.* Consultant report for the World Bank, Washington, DC.

Lodhi, Z. H. 2009. "Ambient Air Quality in Pakistan." Pakistan Environmental Protection Agency (Pak-EPA) http://www.environment.gov.pk/PRO_PDF/AmbientAirQty Pakistan.pdf.

Mansha, M., B. Ghauri, S. Rahman, and A. Amman. 2011. "Characterization and Source Apportionment of Ambient Air Particulate Matter ($PM_{2.5}$) in Karachi." *Science of the Total Environment.* doi: 10.1016/j.scitotenv.2011.10.056.

Mercer. 2010. *Quality of Living Reports–2010 Edition.* http://www.mercer.com/articles /quality-of-living-survey-report-2010.

Muller, N. Z., and R. Mendelsohn. 2007. "Measuring the Damages of Air Pollution in the United States." *Journal of Environmental Economics and Management* 54: 1–14. http:// www.sciencedirect.com/science/article/pii/S0095069607000095.

Murphy, J. J., M. A. Delucchi, D. R. McCubbin, and H. J. Kim. 1999. "The Cost of Crop Damage Caused by Ozone Air Pollution from Motor Vehicles." *Journal of Environmental Management* 55: 273–89. http://www.its.ucdavis.edu/publications/1999/UCD -ITS-RP-99-03.pdf.

OECD (Organisation for Economic Co-operation and Development). 2008. *OECD Environmental Outlook to 2030.* Paris: OECD. http://www.oecd.org/dataoecd /29/33/40200582.pdf.

Ostro, B. 1994. *Estimating the Health Effects of Air Pollution: A Method with an Application to Jakarta.* Policy Research Working Paper Series 1301. Washington, DC: World Bank.

———. 2004. *Outdoor Air Pollution – Assessing the Environmental Burden of Disease at National and Local Levels.* Environmental Burden of Disease Series 5, Geneva, Switzerland: World Health Organization.

Pak-EPA (Pakistan Environmental Protection Agency). 2007. "Ambient Air and Water Quality Investigation in Quetta." Pakistan Environment Programme. http://www .environment.gov.pk/PUB-PDF/Ambient%20AW%20Quetta.pdf.

———. 2010. "Daily $PM_{2.5}$, SO_2, NO_2, O_3, and CO data collected by the Pak-EPA." Unpublished report. Islamabad.

Pope C. A. III, R. T. Burnett, D. Krewski, M. Jerret, Y. Shi, E. E. Calle, and M. J. Thun. 2009. "Cardiovascular Mortality and Exposure to Airborne Fine Particulate Matter and Cigarette Smoke: Shape of the Exposure-Response Relationship." *Circulation* 120: 941–48.

Pope, C. A. III, R. T. Burnett, M. J. Thun, E. E. Calle, D. Krewski, K. Ito, and G. D. Thurston. 2002. "Lung Cancer, Cardiopulmonary Mortality, and Long-Term Exposure to Fine Particulate Air Pollution: Epidemiological Evidence of General Pathophysiological Pathways of Disease." *Journal of the American Medical Association* 287: 1132–41.

Pope, C. A. III, and D. W. Dockery. 2006. "Health Effects of Fine Particulate Air Pollution: Lines that Connect." *Journal of the Air & Waste Management Association* 56: 709–42. http://www.noaca.org/pmhealtheffects.pdf.

Sami, M., A. Waseem, and S. Akbar. 2006. "Quantitative Estimation of Dust Fall and Smoke Particles in Quetta Valley." *Journal of Zhejiang University SCIENCE B* 7 (7): 542–47. http://www.ncbi.nlm.nih.gov/pmc/articles/PMC1500881/?tool=pmcentrez.

Sánchez-Triana, E., S. Enriquez, B. Larsen, and E. Golub. 2014. *Environmental and Climate Change Priorities for the Sindh Province.* Environment and Development Series. Washington, DC: World Bank.

Sindh-EPA. 2010. Daily PM$_{2.5}$ data collected from air quality monitoring network, Karachi. Unpublished.

Strukova, E. 2008. "Punjab Cost of Environmental Damage: An Analysis of Environmental Health." Unpublished draft, World Bank, Washington, DC.

SUPARCO (Pakistan Space and Upper Atmosphere Research Commission). 2004. "Materials on Ambient Air Quality in Major Cities of Pakistan." Karachi.

Surkan, P. J., A. Zhang, F. Trachtenberg, D. B. Daniel, S. McKinlay, and D. C. Bellinger. 2007. "Neuropsychological Function in Children with Blood Lead Levels < 10 µg/dL." *NeuroToxicology* 28 (6): 1170–77.

Ul-Haq, N., M. A. Arain, N. Badar, M. Rasheed, and Z. Haque. 2011. "Drinking Water: A Major Source of Lead Exposure in Karachi, Pakistan." *Eastern Mediterranean Health Journal* 17 (11): 882–86.

Van Dingenen, R., F. J. Dentener, F. Raes, M. C. Krol, L. Emberson, and J. Cofala. 2008. "The Global Impact of Ozone on Agricultural Crop Yields Under Current and Future Air Quality Legislation." *Atmospheric Environment* 43 (3): 604–18. http://www.sciencedirect.com/science?_ob=MImg&_imagekey=B6VH3-4TS6ST0-3-10&_cdi=6055&_user=145269&_pii=S1352231008009424&_origin=&_coverDate=01%2F31%2F2009&_sk=999569996&view=c&wchp=dGLzVzb-zSkWl&md5=ea288891107166e84d5c6fe0e40180cf&ie=/sdarticle.pdf.

Wahid, A. 2006. "Influence of Atmospheric Pollutants on Agriculture in Developing Countries: A Case Study with Three New Wheat Varieties in Pakistan." *Science of the Total Environment* 371: 304–13. http://www.sciencedirect.com/science?_ob=MImg&_imagekey=B6V78-4KKNN42-3-C&_cdi=5836&_user=145269&_pii=S0048969706004566&_origin=&_coverDate=12%2F01%2F2006&_sk=996289998&view=c&wchp=dGLbVzW-zSkWB&md5=8d69d897ea9ab0b4a5950712f29f5b42&ie=/sdarticle.pdf.

WHO (World Health Organization). 2004. *Comparative Quantification of Health Risks: Global and Regional Burden of Disease Attributable to Selected Major Risk Factors.* Geneva, Switzerland: WHO.

———. 2006. "Air Quality Guidelines, Global Update 2005." Regional Office for Europe, Copenhagen, Denmark.

———. 2011. "Violence and Injury Prevention and Disability (VIP)." Pakistan country profile. http://www.who.int/violence_injury_prevention/road_safety_status/country_profiles/pakistan.pdf.

World Bank. 2006. "Pakistan Strategic Country Environmental Assessment." South Asia Environment and Social Development Unit, Washington, DC. http://www.esmap.org/esmap/sites/esmap.org/files/FR275-03_Thailand_Reducing_Emissions_from_Motorcycles_in_Bangkok.pdf.

———. 2008a. *Environmental Health and Child Survival, Epidemiology, Economics, Experiences.* Washington, DC: International Bank for Reconstruction and Development/World Bank. http://siteresources.worldbank.org/INTENVHEA/Resources/9780821372364.pdf?item_id=6981658.

———. 2010a. *World Bank Development Indicators.* http://data.worldbank.org/indicator.

———. 2010b. *Worldwide Governance Indicators.* http://info.worldbank.org/governance/wgi/sc_country.asp.

World Gazetteer. 2011. Population data and statistics. http://www.world-gazatteer.com.

CHAPTER 3

Air Quality Management Institutions

Introduction

The 1997 Pakistan Environmental Protection Act (PEPA) sets the institutional foundation for environmental quality management. National Environmental Quality Standards (NEQS) were first developed in 1993 and later revised in 2000 and 2009. The Pakistan Clean Air Program (PCAP), which was approved in 2005, contained proposals for designing and implementing interventions to achieve ambient air quality objectives. Acute institutional weaknesses and lack of funding have impeded starting up PCAP's implementation. Pakistan urgently needs to intensify its efforts to implement a broad strategy for reducing general urban air pollution, particularly fine particulate matter ($PM_{2.5}$).

In the short term, Pakistan should

- update the air pollution control regulatory framework;
- establish a central apex organization responsible for intersectoral and intergovernmental (national and provincial) coordination;
- build capacity to design and implement air quality management (AQM) policies, including recruiting specialized staff;
- invest in a robust air quality monitoring program; and
- strengthen the environmental capacity of the judiciary bodies.

Air pollution levels in Pakistan's cities are greater than those of Bangkok, Los Angeles, Mexico City, Rome, Santiago, or Tokyo. The differences in concentrations of PM are larger when compared with cities with strong environmental institutions.[1] In developed and developing countries, large metropolises with strong institutions have successfully reduced their ambient concentrations to a level that is lower than even those of small Pakistani cities, such as Gujranwala. Institutional weakness and poor environmental governance help explain such high levels of air pollution in Pakistan's urban centers.

The PEPA defines roles for different organizations, including provincial environmental agencies, and a central enforcement agency. While Pakistan has developed environmental institutions, they are still evolving. The 18th Constitutional Amendment phased out the central-level Ministry of Environment (MoE) responsible for overall policy formulation and coordination, and created the Ministry of Climate Change (MoCC), with a mandate for transboundary pollution issues, climate change issues, and development of national environmental policies. However, the MoCC was downgraded from a ministry to a division in 2013. The existing institutions can serve as the foundations for improved AQM, as long as they are strengthened and reformed based on lessons learned from experiences in Pakistan and elsewhere.

This chapter provides an overview of the current state of Pakistan's air quality regulatory framework. The section on "Pakistan's Air Quality Regulatory Framework" contains a brief introduction to the historical evolution of Pakistan's regulatory framework on air pollution over the period 1993–2013. The section on "Pakistan's Air Quality Regulatory Framework" also discusses the development of the PEPA, the design and implementation of the National Environment Policy of 2005, the PCAP adopted in 2010, and the NEQS. The following section, "Organizational Structure for Air Quality Management," describes the organizational structure for AQM in Pakistan. The section on "Coordination and Decentralization of Air Quality Management Responsibilities" discusses the coordination and decentralization of AQM responsibilities in Pakistan. The following section, "Monitoring Ambient Air Pollution," assesses Pakistan's capacity to measure ambient air pollution and to monitor stationary sources. "The Judiciary and the Enforcement of Regulations" section identifies program areas that need to be strengthened for enforcing air quality regulations, including the capacity needs of the legal courts and, in general, the judiciary system in Pakistan. Finally, the "Conclusions and Recommendations" section proposes interventions to build institutional capacity for AQM in Pakistan.

Pakistan's Air Quality Regulatory Framework

The framework for Pakistan's AQM system dates back to 1993, when the NEQS were developed (table 3.1). On December 6, 1997, Parliament adopted the PEPA, repealing the 1983 Pakistan Environmental Protection Ordinance to provide a comprehensive framework for regulating environmental protection. The Pakistan Environmental Protection Council (PEPC) was created at the federal level, with the Prime Minister of Pakistan as its Chairperson. This council's membership consists of the Provincial Governments, concerned Federal Ministries and Divisions, nongovernmental organizations (NGOs), and the private sector. This was followed by approval of the National Conservation Strategy and appointment of a full-fledged Minister for the Environment (Ghauri 2010, 1).

PEPA is the cornerstone of environmental legislation in Pakistan. The Act established the general conditions, prohibitions, penalties, and enforcement to prevent and control pollution, and to promote sustainable development. The Act

Table 3.1 Development of Pakistan's Norms on Air Pollution, 1983–2009

Act or regulation	Year	Requirement
Pakistan Environmental Protection Ordinance	1983	Established the requirement to prepare an Environmental Impact Assessment for development projects
National Environmental Quality Standards (NEQS)	1993	Issued standards applicable to industrial and municipal liquid effluents and industrial gaseous emissions
Pakistan Environmental Protection Act (PEPA)	1997	Replaced the Pakistan Environmental Protection Ordinance of 1983 Established PEPC, which is responsible for approving the NEQS Gave a formal mandate for the Pak-EPA to propose NEQS and enforce approved standards Enabled the government to levy a pollution charge to those exceeding NEQS and mandated that those who pay it cannot be charged with an offense for exceeding the standards Enabled the government to direct motor vehicles to install pollution control devices, use specified fuels, and undergo prescribed maintenance or testing
Revised National Environmental Quality Standards (NEQS)	1999	Relaxed NEQS, which were considered more stringent than those of other countries in the region, and adjusted them based on Pakistan's conditions and practice in South Asia
National Environmental Action Plan	2001	Introduced unleaded gasoline and reduced sulfur in diesel in Pakistan
National Environment Policy	2005	Set as a priority the establishment of ambient air quality standards, enacting the National Clean Air Act and updating emissions standards for mobile and stationary sources
Pakistan Clean Air Program	2005	A set of interventions designed to attack the primary sources of urban air pollution
National Air Quality Standards	2009	Revised emission standards for all new and in-use vehicles approved by the PEPC
National Air Quality Standards	2010	Establishes NEQS for ambient air quality approved by the PEPC

Note: Pak-EPA = Pakistan Environmental Protection Agency; PEPC = Pakistan Environmental Protection Council.

delineated the responsibilities of the PEPC, Pakistan Environmental Protection Agency (Pak-EPA), and provincial Environmental Protection Agencies (EPAs). It includes provisions for air pollution control.

The NEQS were originally issued under the 1983 Environmental Protection Ordinance. Consultations with major stakeholders began in April 1996. The PEPC approved a revised version of the NEQS in December 1999, and the NEQS became effective in August 2000. The review was justified by the PEPC on grounds that some of the original parameters were more stringent than parameters for other countries in the South Asia region. New standards for exhaust emissions from motor vehicle were approved in 2009.[2]

In 2010, Pak-EPA drafted NEQS for ambient air quality. The NEQS for ambient air cover several major pollutants: (a) Sulfur Dioxide (SO_2); (b) Nitrogen Oxides (NO_x); (c) Ozone (O_3); (d) Suspended Particulate Matter (SPM); (e) Fine Particulate Matter ($PM_{2.5}$); (f) Lead (Pb); and (g) Carbon Monoxide (CO). As required by law, prior to submitting the standards for the review and approval of the PEPC, Pak-EPA published the draft NEQS on its website and requested comments from the public.[3] Both standards were approved by PEPC in a meeting held on March 29, 2010, and the official notifications in the Gazette of Pakistan were made on November 26, 2010 (table 3.2) (Gazette of Pakistan 2010).

Cleaning Pakistan's Air • http://dx.doi.org/10.1596/978-1-4648-0235-5

Table 3.2 Comparison of Pakistan's National Air Quality Standards with WHO, EU, and U.S. Air Quality Guidelines

Pollutants	Time-weighted average	Pakistan ambient air quality standards Effective 2010	Effective 2013	WHO air quality guidelines[a]	EU ambient air quality standards	U.S. ambient air quality standards
Suspended Particulate Matter (SPM)	Annual average[b]	400 μg/m³	360 μg/m³	n.a.	n.a.	n.a.
	24 h[c]	550 μg/m³	500 μg/m³	n.a.	n.a.	n.a.
Particulate Matter (PM$_{10}$)	Annual average[b]	200 μg/m³	120 μg/m³	20 μg/m³	40 μg/m³	n.a.
	24 h[c]	250 μg/m³	150 μg/m³	50 μg/m³	50 μg/m³	150 μg/m³
Particulate Matter (PM$_{2.5}$)	Annual average[b]	25 μg/m³	15 μg/m³	10 μg/m³	25 μg/m³	15 μg/m³
	24 h[c]	40 μg/m³	35 μg/m³	25 μg/m³	n.a.	35 μg/m³
	1 h	25 μg/m³	15 μg/m³	1 μg/m³	n.a.	n.a.
Lead (Pb)	Annual average[b]	1.5 μg/m³	1 μg/m³	0.5 μg/m³	0.5 μg/m³	n.a.
	24 h[c]	2 μg/m³	1.5 μg/m³	n.a.	n.a.	n.a.
Sulfur Dioxide (SO$_2$)	Annual average[b]	80 μg/m³	80 μg/m³	n.a.	n.a.	85.8 μg/m³
	24 h[c]	120 μg/m³	120 μg/m³	20 μg/m³	125 μg/m³	n.a.
Nitrogen Dioxide (NO$_2$)	Annual average[b]	40 μg/m³	40 μg/m³	40 μg/m³	40 μg/m³	100 μg/m³
	24 h[c]	80 μg/m³	80 μg/m³	n.a.	200 μg/m³	188 85.8 μg/m³
Nitric Oxide (NO)	Annual average[b]	40 μg/m³	40 μg/m³	n.a.	n.a.	n.a.
	24 h[c]	40 μg/m³	40 μg/m³	n.a.	n.a.	n.a.
Carbon Monoxide (CO)	8 h[c]	5 mg/m³	5 mg/m³	n.a.	10 mg/m³	10 mg/m³
	1 h	10 mg/m³	10 mg/m³	n.a.	n.a.	40 μg/m³

Sources: Colbeck, Zaheer, and Zulfiqar 2010, 60; Gazette of Pakistan 2010; Ghauri 2010.
Note: EU = European Union, n.a. = not applicable, PM$_{2.5}$ = particulate matter of less than 2.5 microns, PM$_{10}$ = particulate matter of less than 10 microns, WHO = World Health Organization.
a. WHO guidelines only given if the averaging period is identical.
b. Annual arithmetic mean of minimum 104 measurements in a year, taken twice a week every 24/h at uniform intervals.
c. Twenty-four-hour/8-h values should be met 98% of the year. It may be exceeded 2% of the time but not on consecutive days.

Pakistan pollution charges established by PEPA 1997 provide an opportunity to implement the polluter pays principle (PPP). PEPA's Section 11(2) on the levy of pollution charges states that the Federal Government may levy a pollution charge on any person who discharges, emits, or allows the discharge of emission of any effluent or waste of air pollutants or noise in an amount, concentration, or level that exceeds the NEQS. PCAP's main objective is to reduce the health and economic impacts of air pollution by implementing a number of short-term and long-term measures that require action at all levels of government (table 3.3).

Pakistan lags behind countries like Bangladesh, China, India, and Sri Lanka in the implementation of AQM interventions. For example, Bangladesh, China, India, and Sri Lanka have implemented vehicle emissions testing pilots. Some of these countries have also made significant progress in establishing systematic air quality measurement capacity. In India, the national air quality monitoring network has operations in 115 cities; in China, 2,289 monitoring stations operate at all administrative levels and the AQM network is expected

Table 3.3 PCAP's Proposed Activities to Address Air Pollution in Pakistan

Short-term activities	Responsible agencies	Long-term measures	Responsible agencies
General air quality management			
Baseline data collection on ambient air quality using fixed and mobile laboratories	Federal and Provincial EPAs	Creation of public awareness and education	MoE and Provincial Environment Department
Launch of effective awareness campaign against smoke-emitting vehicles	Provincial Governments	Setting up continuous monitoring stations in cities to record pollution levels in ambient air	MoE and Provincial Government
Reducing emissions from mobile sources			
Stop import and local manufacturing of two-stroke vehicles	Ministry of Commerce and Ministry of Industry	Improvement of energy efficiency in vehicles	MoE
Restriction on conversion of vehicles from gasoline engine to secondhand diesel engines; launch effective awareness campaign against smoke-emitting vehicles	Provincial Governments	Introduction of low-sulfur diesel and furnace oil and promotion of alternative fuels—such as compressed natural gas (CNG), liquid petroleum gas (LPG), and mixed fuels—in the country	Ministry of Petroleum and Natural Resources (MoPNR)
High pollution spots in cities may be identified and controlled through better traffic management, such as establishment of rapid mass transit and traffic-free zones	Provincial Governments	Review Motor Vehicle Ordinance to provide for inspection of private vehicles	Federal and Provincial Governments
Capacity building of motor vehicle examiners	Provincial Governments	Establish vehicle inspection centers	Ministry of Communication and Provincial Government
Regular checking of quality of fuel and lubricating oils sold in the market	MoPNR	Identify pollution control devices and additives for vehicles and encourage their use	MoPNR
Phasing out of two-stroke and diesel-run public service vehicles	Federal and Provincial Governments		
Giving tariff preference to CNG-driven buses	Ministry of Industries and Ministry of Finance (MoF)		
Adoption of fiscal incentives and a financing mechanism to provide resources to transporters	Ministry of Communication and Provincial Government		
Establishment of environmental squad of traffic police in all major cities to control visible smoke	Provincial Governments		

table continues next page

Table 3.3 PCAP's Proposed Activities to Address Air Pollution in Pakistan (continued)

Short-term activities	Responsible agencies	Long-term measures	Responsible agencies
Reducing emissions from stationary sources			
Covering of buildings and sites during renovation and construction to avoid air pollution	Provincial Governments	Promotion of waste minimization, waste exchange, and pollution control technology in industries	Federal and Provincial EPAs, Federation of Pakistan Chamber of Commerce and Industries, and Ministry of Industries and Production
Reducing emissions from area sources (open burning) and dust			
		Proper disposal of solid waste in cities/provinces	Capital Development Authority and Provincial Governments
		Block tree plantation in cities, forestation in deserts, and sand dune stabilization	MoE and Provincial Forest Department
		Paving of shoulders along roads	Ministry of Communication and Provincial Government

Source: Pak-EPA 2005.
Note: EPAs = Environmental Protection Agencies; MoE = Ministry of Environment.

to become automatic in the short term. A review of interventions in the region and across the world confirms the gap in implementation of AQM interventions. International good practices could serve as benchmarks for Pakistan's AQM program.

Organizational Structure for Air Quality Management

The PEPA assigns leadership responsibilities for environmental management to the PEPC. The Prime Minister of Pakistan heads the Council, and the Federal MCC serves as the Vice-Chairman. The council's membership is comprised of multiple stakeholders, including (a) Chief Ministers of the four provinces; (b) Provincial Environmental Ministers; (c) 35 ex-officio representatives (from sectors that include industry, technical and professional associations, trade unions, and NGOs); and (d) the head of the Climate Change Division. Major functions of PEPC are to (a) supervise the implementation of PEPA 1997, (b) approve and supervise the implementation of national environmental policies, (c) approve NEQS, (d) provide guidelines for the protection and conservation of natural resources and habitats, (e) integrate sustainable development in national development plans and policies, and (f) instruct relevant institutions to execute sustainable development and research projects. Unfortunately, the Council meets irregularly. From 2005 to 2010, the Council met only once (Pakistan Today 2011).

The MoE was established as a full-fledged ministry in 2002. Previously, it was part of the Ministry of Local Government and Rural Development. The MoE was established under the Federal Government Order (not covered by PEPA 1997), and it was second in the hierarchy of environmental institutions. After the adoption of the 2010 18th Constitutional Amendment, the MoE was absorbed by the Ministry of National Disaster Management, which became the Ministry of Climate Change in April 2012. However, the ministry was downgraded to a division in June 2013.

Coordination and Decentralization of Air Quality Management Responsibilities

The physical boundaries of air pollution rarely coincide with those of existing administrative or political boundaries (districts, municipalities, and provinces). As a result, the need for intergovernmental and intersectoral coordination emerges, nationally and internationally, vertically and horizontally.

Systematized mechanisms for intersectoral coordination to tackle crosscutting issues and harmonize common interventions have not been set in Pakistan. In spite of all the directives contained in the regulations stressing the importance of coordination among concerned agencies, by June of 2014, no formal mechanisms existed for agencies involved in environmental management to participate in a consultative process with other provincial or sectoral agencies for priority-setting, design and implementation of interventions, monitoring, and evaluation of effectiveness. Intersectoral coordination for the oversight of crosscutting issues is also nonexistent. Some attempts have been made to establish focal points within other non-environment ministries, but interactions among these focal points have not yet been institutionalized.

Several sectoral ministries are important players in the design and implementation of AQM policies. The Ministry of Petroleum and Natural Resources (MoPNR) is responsible for combating adulteration of fuel and increasing standards, particularly by lowering sulfur content for fuel refined in the country. The Ministry of Industry and Special Initiatives (MoI) is responsible for regulating the types of vehicles that could be imported, potentially constraining imports of high-polluting vehicles at the gate. The MoI is also responsible for measures aimed at modernizing the fleet of public service vehicles and scrapping older vehicles. The Ministry of Finance (MoF) and MoPNR are responsible for fuel pricing and subsidies. The Ministry of Energy (MoEn) is responsible for clean fuel imports and encouraging the use of CNG in vehicles. Finally, the Ministry of Agriculture (MoA) is responsible for regulating burning of sugarcane fields and agricultural waste.

The 18th Constitutional Amendment devolved responsibilities for environmental management to subnational governments. Since the adoption of the 18th Constitutional Amendment, provincial governments have taken over environmental management responsibilities in an ad hoc manner. In Punjab, for example, District Environment Officers have been appointed in most districts. However, in the other three provinces, the environment departments have set up regional offices. While decentralization of environmental management responsibilities offers a number of benefits, including the capacity to respond more effectively to local priorities, there are also significant tradeoffs and risks. For example, unequal definition and enforcement of environmental standards, as well as differences in the capacity of environmental agencies, could lead to more severe environmental degradation in different parts of the country.

Most countries in the world currently have an apex central environmental ministry or agency with a number of technical and action-oriented agencies designating and implementing public policies, and enforcing regulations. Since environmental problems are typically felt locally, provinces and municipalities are often in a better position to address environmental problems, and thus would achieve superior outcomes if given the freedom to choose the most appropriate policies and instruments. This is the immediate rationale for decentralizing environmental management. However, without proper coordination, decentralization eventually leads to significant differences in environmental quality across regions. In Pakistan, delegation of environmental functions from the federal government to provincial governments is comprehensive and has empowered provincial EPAs to take care of most environmental issues in the provinces. After the delegation of enforcement functions to the provincial governments, Pak-EPA's main responsibilities are limited to assisting provincial governments in the formulation of rules and regulations under PEPA 1997.

Coordination is critical for successful decentralization: the transfer of responsibilities has the potential to make the coordination of national policies difficult, particularly in federalist systems. In Brazil, Mexico, and the United States, local environmental agencies were given substantial freedom to determine the way in which environmental standards are met. However, all these nations retained an apex-level body, located at the federal level, to make environment policy and manage coordination between states and provinces. Even countries that have achieved good results from decentralization, such as Brazil and Canada, have found that some environmental management functions cannot be successfully decentralized. These are kept as central responsibilities because failure to do so has been found to be potentially harmful to the environment and the population.

Specifically, the responsibilities that the central government tends to maintain, regardless of the level of decentralization, include

- Design and enactment of national environmental policies and standards that provide consistency of rules and regulations;
- Transboundary issues, including representing countries at international negotiations and in international conventions and initiatives, such as the biodiversity or the climate change conventions;

- Coordination of regional agencies, including collaboration and sharing of good practices, and monitoring and evaluation of environmental programs that impact multiple regions;
- Granting environmental licenses and permits for activities that affect the environment in more than one subnational area; and
- Research related to climate change, biodiversity, or water issues, such as glacial melting and other problems that affect multiple provinces and subnational areas, as well as multiple countries.

Some functions of environmental management have been found to be managed more efficiently by subnational provincial environmental agencies. Specifically,

- Enforcement of national regulations and processing of tasks associated with environmental regulations, such as permits and fines for infractions;
- Design and implementation of action plans and strategies to meet national environmental standards at the municipal, regional, and provincial levels;
- Promotion and facilitation of public participation in order to foster greater political and cultural representation, and transparency of the decision-making process;
- Monitoring and dissemination of information, in accordance with national standards, on environmental quality, such as air and water quality; and
- Monitoring of local environmental concerns and dissemination of information to constituents and stakeholders.

Still, there are a number of potentially negative consequences of decentralization of environmental management, including

- Provincial and local governments may not set environmental standards high enough or may not wish to properly enforce them;
- It may become problematic to take environmental actions on a required scale that, for the most significant problems, is generally larger than the territory of a municipality;
- Local governments find it difficult to control problems that result in distant impacts: for example, an upstream municipality may have no incentive to curtail water pollution if those who suffer are residents of provinces located downstream;
- Services may be delivered less effectively and less efficiently in some areas of the country because of weak administrative or technical capacity at local levels; and
- Coordination of national policies may become more complex, stabilization policies more difficult to implement, and the levels and composition of overall public expenditures and public debt may be destabilized.

Monitoring Ambient Air Pollution

The provincial EPAs and the Pak-EPA are in charge of monitoring air pollution in Pakistan. From 2006 to 2009, the Japanese International Cooperation Agency (JICA) assisted the GoP in the design and installation of an air quality monitoring network of measurement stations that included (a) fixed and mobile Air Monitoring Stations in five major cities of Pakistan (Islamabad, Karachi, Lahore, Peshawar, and Quetta); (b) a data center; and (c) a central laboratory. The monitoring units in the provinces were managed and operated by the provincial EPAs. Actual operation was initially carried out by consultants hired, trained, and paid by the Japanese partners, and after the first year of operation, the EPAs were expected to assume all costs related to the monitoring work. In mid-2012, the project shut down as Japanese support phased out and EPAs had not assumed the operation and maintenance costs of the air quality monitoring network. Only the Punjab EPA had developed a project that aimed at continuing with the air quality monitoring efforts started with Japanese assistance.

Administrative and budget problems have led to inadequate operation and maintenance of the air quality monitoring network. Automated instruments were used in Lahore, Karachi, Quetta, Peshawar, and Islamabad, while manual, or manual and automated combinations were used in other areas. The number and location of the monitors had been criticized because they excluded "hot spots" where air quality is believed to be particularly poor. Various technical problems were reported related to the interrupted power supply and difficulties in maintaining the automated electronic instruments. Provincial EPAs faced difficulties retaining the technicians trained in the proper operation of the equipment under the JICA program, as well as calibrating and interpreting the data generated by the stations. This led to considerable downtime and out-of-calibration readings. Furthermore, the data that were collected were neither analyzed nor disclosed.

Thermal inversions that happen across much of Pakistan from December to March lower the mixing height and result in high pollutant concentrations, especially under stable atmospheric conditions. SPM contributes to the formation of ground fog that prevails over much of the Indo-Gangetic Plains during the winter months (Faiz 2011). In addition, sunny and stable weather conditions lead to high concentration of pollutants in the atmosphere (Sami, Waseem, and Akbar 2006). Due to high temperatures in summer (40–50°C), fine dust rises up with the hot air and forms "dust clouds" and haze over many cities of southern Punjab and upper Sindh. Dust storms generated from deserts (Thal, Cholistan, and Thar), particularly during the summer, adversely affect visibility in the cities of Punjab and Sindh (Hussain, Mir, and Azfal 2005).

The concentration of $PM_{2.5}$ is being monitored infrequently. Although several major pollutants, including O_3 and course particulate matter (PM_{10}), are of concern in urban areas, the most serious health effects are caused by $PM_{2.5}$. Recent data show that $PM_{2.5}$ poses the greatest risk of causing or exacerbating lung and heart problems. In addition, Pb contamination is known to be severe in Pakistan because of its accumulation in people's blood (World Bank 2006). However, the specific sources of Pb and other toxic substances are not well known.

The Judiciary and the Enforcement of Regulations

In Pakistan, the judiciary has played an increasingly important role in the enforcement of environmental laws, and should continue to be strengthened through continued support for both judges and advocates. When regulatory avenues for environmental enforcement fail, the judicial system is often the only other recourse for resolving environmental conflicts. An independent judiciary and judicial process enhances implementation, development, and enforcement of regulations bearing on air pollution control. The Supreme Court of Pakistan has considered several cases regarding the degradation of the environment and the protection of a clean environment, and it has concluded that the right to a clean environment is a fundamental right of all citizens of Pakistan, covered by the right to life and right to dignity under Articles 9 and 14 of the Constitution.

The High Courts in the provinces have also intervened and rendered decisions affecting future environmental management. One example of court policy intervention led to the establishment of the Lahore Clean Air Commission. The Lahore High Court appointed the Commission to develop and submit a report on feasible and specific solutions and measures for monitoring, controlling, and improving vehicular air pollution in the city of Lahore.

Section 20 of the PEPA authorized the federal government to establish as many environmental tribunals (ETs) as it considered necessary and specify the territorial limits or class of cases under which each of them could exercise jurisdiction. According to the PEPA, the ETs were to be staffed by Environmental Magistrates appointed, from among senior civil judges, by the federal and provincial governments. The ETs were empowered to sentence repeat offenders to up to two years imprisonment and to order the permanent closure of a factory. Two ETs were created under PEPA, based in Karachi and Lahore, with jurisdiction over other provinces and areas. The ETs' performance was hampered by a number of constraints, including constant suspension of activities, legal loopholes, and lack of clear mechanisms to collect imposed sanctions.

After the 18th Constitutional Amendment's devolution of environmental issues to the provincial governments, the Government of Punjab adopted new rules for the ET of Lahore. The devolution of this responsibility to provincial governments thus opens the opportunity to strengthen ETs and create a more solid legal framework that would result in adequate sanctions being imposed on violators of environmental laws. Provincial governments should consider building the capacity of the Judiciary and ETs. The Courts should also establish mechanisms to monitor the implementation of court orders either through the establishment of judicially appointed oversight committees or through judicially mandated reporting requirements. By establishing a constitutional right to a clean environment and demonstrating a willingness to address matters of environmental policy, the courts have empowered citizens with legal standing to enforce environmental laws through administrative and judicial proceedings. While this right has been established by the courts, there are no citizen suit provisions in the enabling environmental statutes. Existing and future laws should explicitly provide for citizen enforcement. Public interest advocacy is a powerful force for improvements in environmental management and should be supported through environmental law associations and the establishment of environmental law clinics at universities.

Conclusions and Recommendations

Pakistan's AQM framework has evolved rapidly since the first NEQS were adopted in 1993. During this time, Pakistan developed some of the basic pillars that can support efforts to improve air quality in the country's urban areas. However, achieving reductions in air pollution will require a strong political commitment, dedicated resources, and building capacity at different levels. Following is a brief description of key actions that the GoP should consider adopting in the short term.

Consolidating the air pollution control regulatory framework is indispensable to achieve improved urban air quality. The NEQS for ambient air quality published at the end of 2010 provide an important first step in establishing maximum limits that are consistent with the findings of research documenting the health impacts of exposure to different air pollutants. However, the new NEQS still allow for concentrations of some pollutants at levels that are higher than those recommended by the World Health Organization. A different, but also significant challenge, will be enforcing the new standards, including compliance with the different deadlines.

Pollution charges could be used in the short term as an efficient mechanism to reduce pollution emissions from different sources. The provisions in PEPA on the establishment of pollution charges, as well as the detailed formulas that were developed in 2001 for reporting and paying pollution charges, can provide the formal support for the establishment of the charges.

As with any other instrument established by law, the pollution charges will not be effective if they are not strictly enforced. In the medium term, the GoP might consider additional economic instruments to promote the use of cleaner vehicles. These could include taxes based on fuel efficiency standards, with higher taxes imposed on less fuel-efficient vehicles. In the long term, taxes could be introduced on hydrocarbons in proportion to their greenhouse gas (GHG) emissions. Further studies should be carried out to assess the distributional impacts of these measures and propose mechanisms to mitigate any regressive effects.

The Pak-EPA has been leading the design and implementation of the PCAP. In the area of AQM, the Pak-EPA and the provincial environmental agencies have initiated the identification of interventions envisioned by the PCAP. In the short term, these agencies have agreed on the need to redesign their organizational structure, prioritize and elaborate pollution control interventions, facilitate financing from different sources, and involve public participation in the design and implementation of AQM (Ministry of Environment 2010).

The organizational restructuring proposed by Pak-EPA includes establishing units specialized in AQM at the national and provincial levels. The Pak-EPA would take over responsibilities for coordinating, designing, and implementing air quality policies. Technical cells at the national and provincial agencies would be responsible for monitoring ambient air quality, and mobile, stationary, and diffuse emissions. Additionally, other technical cells would be responsible for regulatory enforcement and compliance. Each of these cells should be provided with three to five specialists with specific technical expertise. These units would be responsible not only for the effective planning and implementation of air quality standards in the territory where the national and provincial environmental organizations are directly competent, but would also play a key role in coordinating work of the same nature carried out by all the agencies that comprise the institutional network of AQM.

Improving air quality in Pakistan will require the establishment of a central apex organization responsible for intersectoral and intergovernmental (national and provincial) coordination. Experiences from around the world, including both developed and developing countries, indicate that the existence of an apex organization is necessary to ensure coordination among different economic sectors, as well as across different levels of government.

Enhancement of interagency and intersectoral coordination can contribute to government effectiveness in the area of AQM. Ghauri (2010) and Parel (2010) suggest that the role of other ministries in AQM needs to be coordinated by means of institutionalized mechanisms, but it is highly recommended that an apex body take a leading role in institutional coordination for implementing AQM across different administrative levels through systematized agreements and coordinated actions.

Recent modifications in Pakistan's Constitution have devolved major responsibilities for environmental management to subnational governments, which will have significant implications for AQM. Since the adoption of the 18th Constitutional Amendment, provincial governments have devolved environmental management responsibilities in an ad hoc manner. While decentralization of environmental management responsibilities offers a number of benefits, including the capacity to respond more effectively to local priorities, there are also significant tradeoffs and risks. For example, unequal definition and enforcement of environmental standards, as well as differences in the capacity of environmental agencies, could lead to more severe environmental degradation in different parts of the country.

Even countries that use decentralized environmental management approaches maintain responsibilities impossible to delegate to regional entities. Some of the responsibilities are maintained centrally for reasons of efficiency, but others are kept as central responsibilities because failure to do so has been determined to be potentially harmful to the environment and the population. Specifically, the responsibilities that tend to be maintained by the central government, regardless of the level of decentralization, deal with (a) enacting environmental standards and policies; (b) national and international transboundary issues, including international agreements; (c) coordination between local governments; and (d) research into environmental issues, such as climate change adaptation and mitigation (Environmental Law Institute 2010).

Uncontrolled decentralization can be very ineffective, as local interests are in a better position to degrade and deplete the resources faster and more efficiently. Not all environmental functions can or should be financed and managed in a decentralized fashion, and even when national governments decentralize respon- sibilities, they must retain important policy and supervisory functions. In addi- tion to political will, environmental decentralization should be implemented in light of the size of the country, the nature of the most serious environmental problems, and the actual strengths and weaknesses of public and private sector organizations of different levels.

Coordination is also critical for successful decentralization. Decentralization efforts may fail tremendously without a reasonable level of supervision and monitoring by central governments, and solid coordination between agencies. Even when local capacity is strong, the transfer of responsibilities may make the coordination of national policies difficult, particularly in federative systems. Even in Canada, a country with strong formal mechanisms for coordination, the underlying cause of good coordination is the existence on an apex agency at the federal level (Margulis and Vetleseter 1999; World Bank 2010).

A top short-term priority to strengthen AQM consists of building capacity to design and implement AQM policies, including recruiting specialized staff. The organizational restructuring proposed by Pak-EPA requires specialists who can carry out a range of actions, including monitoring, enforcement, and planning. Recruiting the staff with the necessary expertise and background will be paramount to ensure that technical cells, as well as any apex organiza- tion that may be created, can fulfill their responsibilities. However, in some instances, outsourcing of these functions to specialized firms may be more efficient.

Data regarding urban air quality in Pakistan are scarce, dispersed, and not fully reliable. The latest available data were collected by means of a recently installed Environmental Monitoring System. However, restricted budget support for the JICA-funded stations affected the availability of trained personal, maintenance parts, and consumables. EPAs have not assumed the costs to maintain the operation of the monitoring network, which consequently suspended its opera- tions. While the program operated, there was no auditing of the monitoring and evaluation program to verify the accuracy of results. Furthermore, the data that were generated were not used to identify or prioritize interventions. Some other problems, such as the high concentration of desert dust during summer time, contributed to the misinterpretation of collected data and their attribution to different polluting sources. Furthermore, the concentration of $PM_{2.5}$, one of the most harmful pollutants for human health, is infrequently monitored and needs to be included in a continuous monitoring regime.

The technical capacity of existing institutions should be strengthened by developing partnerships with research centers to conduct applied research, improve local and regional models, and create centers of excellence in the country on AQM. In addition, air pollution is extremely dependent upon meteorology, and the national and regional meteorological institutes should be strengthened and linked to air pollution entities.

Investing in a robust air quality monitoring program is essential to understand fully the risks generated by air pollution and the effectiveness of government interventions carried out to address it.

Currently, there is almost no systematic monitoring of ambient air quality in large urban centers. The monitoring equipment and assistance from JICA established a basic capability. However, ideally, the current monitoring program would be expanded and ongoing training and budget would be provided to maintain and utilize the equipment. Air quality monitoring is essential in order to estimate health problems and other consequences of poor air quality.

Continuous monitoring is needed to be able to identify the changes in air quality over time. Monitoring is also necessary in order to frame campaigns and build constituencies for targeted interventions. Ensuring the sustainability of monitoring efforts requires establishing a constant stream of resources to pay for technicians' salaries, equipment maintenance, and consumables. Once a reliable monitoring network is in place, modeling efforts should be undertaken to identify the contribution of different natural and anthropogenic sources of air pollutants in urban areas.

In the short term, the GoP should establish a reliable air quality monitoring network focusing on pollutants such as $PM_{2.5}$, SO_2, nitrate (NO_3), and Pb and other toxic substances; develop a detailed mobile source emissions inventory; and establish an inventory of industrial sources, focusing on key polluters and evolving to include small and medium enterprises. In the medium term, the GoP should complement these efforts by establishing a centralized depository. This entity would review and analyze data collected by the air quality monitoring network from across the country, as well as carrying out modeling and speciation efforts to assess the present and future contributions of mobile, stationary, nonpoint, and natural sources of key pollutants, as well as their composition.

Strengthening enforcement of air quality regulations and the capacity of the legal courts and judicial bodies should not be postponed any longer. Enforcement of air quality regulations could be a responsibility of the new technical cells proposed by Pak-EPA. In addition, the judicial system should be strengthened to act as an additional recourse for resolving environmental conflicts when enforcement fails.

Table 3.4 Government Interventions to Strengthen Institutional EPA Efforts

Action	Relevant regional examples
Upgrade laws and regulations to eliminate ambiguities	
Develop institutional capacity to target and enforce pollution monitoring and abatement in high polluting industries and companies.	*Bangladesh*: 1997 and earlier legislation provides standards for industrial emissions. *China*: "Top 1,000 Enterprises Program" enforces energy efficiency on the largest firms (using 1/3 of the country's energy), retiring inefficient power plants, and closing inefficient industrial plants. Conservation law promotes end-use energy efficiency, and introduces export taxes on energy-intensive products. There are two types of emission standards for stationary pollution sources: (a) general 'integrated emissions standard of air pollutants,' and (b) those for particular industry types (boilers, thermal power plants, kilns, furnaces, coke ovens, and cement plants), plus 'total emission control' ceilings. Pollution levies have been assessed since 1982. *India*: Key actions include retiring inefficient coal plants by 2012 and mandating use of washed coal at urban power plants to reduce ash. Main instruments include the Energy Conservation Act and Bureau of Energy Efficiency, Energy Conservation Building Code, and National Tariff Policy (higher tariffs for large consumers). The Central Pollution Control Board established maximum limits for different pollutants for many categories of industries—notified by the government of India under EPA of 1986. Submission of an environmental statement by polluting units is mandatory. *Sri Lanka*: Clean Air 2015 Action Plan.

At least one ET should be created in each province and in Islamabad; this would imply creating three new ETs in addition to those established in Karachi and Lahore. In addition, existing and future laws should explicitly provide for citizen enforcement. Finally, public interest advocacy is a powerful force for improvements in environmental management and should be supported through environmental law associations and the establishment of environmental law clinics at universities.

There are several regional examples of interventions that the GoP might use to inform its institutional strengthening efforts. Table 3.4 summarizes these examples.

Notes

1. According to North (1990, 3), "Institutions are the rules of the game in a society or, more formally, are the humanly devised constraints that shape human interaction." Institutions can be formal (such as laws) and informal (including conventions and codes of behavior). North (1990, 3) further states "… the formal and informal rules and the type and effectiveness of enforcement shape the whole character of the game." Organizations, which are different from institutions, can be thought of as the players of the game.

2. NEQS for Motor Vehicle Exhaust and Noise. These NEQS (also known in Pakistan as Euro 2 Standards) replace Annex III of the Notification S.R.O. 742 (I) 93, dated August 24, 1993. The S.R.O. 72 (KE)/2009 introduces a set of emission standards for all new and in-use vehicles. Those categories are subdivided into diesel (light and heavy) and gasoline (petrol)-powered vehicles (passenger cars, light commercial vehicles, rickshaws, and motorcycles). For example, for heavy diesel engines and large goods vehicles (both locally manufactured and imported), the standard for PM

is 0.15 g/kWh, to be enforced after July 1, 2012 (the standard does not define if this value refers to course particulate matter (PM_{10}), or fine particulate matter ($PM_{2.5}$)). No reference is made to standards for sulfur dioxide (SO_2). The schedule of implementation includes three deadlines: immediate; July 1, 2009; and July 1, 2012. The smoke, carbon monoxide (CO), and noise standards for in-use vehicles were to be effective immediately after the regulation was adopted. The S.R.O. 1062 (I)/2010 established NEQS for ambient air quality. Its schedule of implementation includes two deadlines: July 1, 2010, and January 1, 2013.

3. According to Faiz (2011), "… the emission standards for new vehicles are generally different from in-use vehicles as the needs and test procedures are different. Euro 2 standards require 3-way catalysts. For heavy-duty diesel vehicles, PM and NO_x standards are specified together as there is tradeoff between these two in terms of engine optimization."

References

Colbeck, I., A. Zaheer, and A. Zulfiqar. 2010. "The State of Ambient Air Quality in Pakistan: A Review." *Environmental Science and Pollution Research* 17: 49–63. http://www.springerlink.com/content/f718jn535422j0wh/fulltext.pdf.

Environmental Law Institute. 2010. *India 2030: Vision for an Environmentally Sustainable Future. Best Practices Analysis of Environmental Protection Authorities in Federal States.* Study commissioned by the World Bank. Washington, DC: World Bank. http://www.eli.org/pdf/india2030.pdf.

Faiz, A. 2011. Comments and Edits on Draft Air Quality Management Report. E-mail messages from June 19, 26, 29, and 30. Personal communication with authors.

Gazette of Pakistan. 2010. *Statutory Notification S.R.O. 1062(I)/2010.* Islamabad: Ministry of Environment of the Government of Pakistan.

Ghauri, B. 2010. *Institutional Analysis of Air Quality Management in Urban Pakistan.* Consulting report commissioned by the World Bank, Washington, DC.

Hussain, A., H. Mir, and M. Afzal. 2005. "Analysis of Dust Storms Frequency over Pakistan during 1961–2000". *Pakistan Journal of Meteorology* 2 (3): 49–68.

Margulis, S., and T. Vetleseter. 1999. *Environmental Capacity Building: A Review of the World Bank's Portfolio.* Environment Department Paper 68, Pollution Management Series, Washington, DC: World Bank.

Ministry of Environment. 2010. "Economic Survey of Pakistan 2009." Ministry of Textile Industry. http://www.finance.gov.pk/survey_0910.html. Accessed April 8, 2010.

North, D. C. 1990. *Institutions, Institutional Change and Economic Performance.* Cambridge, U.K.: Cambridge University Press.

Pak-EPA. 2005. "Pakistan Clean Air Program." Islamabad. http://www.environment.gov.pk/NEP/PCAPFinal.pdf.

Pakistan Today. 2011. "Toothless Pakistan Environmental Protection Council." March 26. http://www.pakistantoday.com.pk/2011/03/26/national/toothless-pakistan-environmental-protection-council.

Parel, C. 2010. Unpublished consultant report for the World Bank, Washington, DC.

Sami, M., A. Waseem, and S. Akbar. 2006. "Quantitative Estimation of Dust Fall and Smoke Particles in Quetta Valley." *Journal of Zhejiang University SCIENCE B* 7 (7): 542–47. http://www.ncbi.nlm.nih.gov/pmc/articles/PMC1500881/?tool=pmcentrez.

World Bank. 2006. *Pakistan Strategic Country Environmental Assessment.* South Asia Environment and Social Development Unit. Washington, DC. http://www.esmap.org /esmap/sites/esmap.org/files/FR275-03_Thailand_Reducing_Emissions_from _Motorcycles_in_Bangkok.pdf.

———. 2010. *Worldwide Governance Indicators.* http://info.worldbank.org/governance /wgi/sc_country.asp.

CHAPTER 4

Mobile Sources

Among mobile sources in Pakistan, the largest contribution of air emissions, particularly coarse and fine particulate matter (PM_{10} and $PM_{2.5}$), comes from incomplete combustion of fuel by diesel-powered vehicles including trucks, buses, and auto-rickshaws. Economically efficient interventions for reducing air pollution from mobile sources include, among others

- Reducing sulfur content in diesel to 500 parts per million (ppm) in the short term and to 50 ppm in the medium term
- Continuing to convert diesel-fueled minibuses and city delivery vans to compressed natural gas (CNG) and install diesel oxidation catalysts (DOC) on existing large buses and trucks used in the city
- Sustaining the introduction of new CNG full-size buses, as diesel particulate filters (DPFs) cannot be used with 500-ppm sulfur in diesel
- Converting existing two-stroke rickshaws to four-stroke CNG engines
- Banning new two-stroke motorcycles and rickshaws and exploring options to control particulate matter (PM) emissions from in-use two-stroke motorcycles
- Introducing low-sulfur fuel oil (1% sulfur) to major users located in urban centers
- Controlling emissions from large point sources
- Restricting use of CNG in spark-ignition automobiles and introducing Euro 2 and higher standards on locally assembled and imported private automobiles and light-duty utility vehicles.

Actions that the government is already pursuing, to various degrees, include conversion of rickshaws to CNG, banning two-stroke vehicles, selectively introducing low-sulfur fuel oil, and banning waste burning within Karachi.

Introduction

Based on available data, mobile sources are the largest contributor of air pollutants, particularly coarse particulate matter (PM_{10}), in large urban centers in Pakistan.[1] Within the past two decades, the number of vehicles has increased more than fivefold from 2.1 million to about 10.6 million. Diesel with high-sulfur content used by buses, trucks, and two and three wheelers generates most of the mobile-source emissions of PM. These vehicles contribute more than 70% of the fine particulate matter ($PM_{2.5}$) and PM_{10} emissions in large urban centers. Two and three wheelers account for 56% of Pakistan's motor vehicle fleet. In Karachi, approximately 32–52% of exhaust hydrocarbon and PM emissions from all motor vehicle activity are attributable to two wheelers. Underpowered engines, the result of overloading—carrying much more weight than the maximum specified by the vehicle manufacturer—further increase the rate of pollution (Ghauri 2010; Gwilliam, Kojima, and Johnson 2004).

This chapter analyzes options to control air pollution from mobile sources in Karachi. The section on "Ambient Air Quality and Exposed Population in Karachi" describes the conditions and effects of PM pollution in Karachi, which explain why a section on Karachi is merited in this book. The "Lead Exposure" section discusses blood lead levels (BLL) and their effects on children under 5 years, including neuropsychological effects. The section on "Neuropsychological Effects in Children under 5 Years" provides economic estimates of the costs of PM pollution and lead exposure in Karachi. The "Social Cost of Health Effects" section discusses the social cost of air pollution's effects on human health. The section on "Interventions to Improve Air Quality" analyzes a series of interventions that could be adopted to improve air quality, and explains that the benefits that each of these interventions would accrue exceed the associated costs. Finally, the "Conclusions and Recommendations" section provides the chapter's conclusions and recommendations.

Ambient Air Quality and Exposed Population in Karachi

PM is the outdoor air pollutant that globally is associated with the largest health effects. The World Health Organization (WHO) recently reduced its guideline limits to an annual average ambient concentration of 10 micrograms per cubic meter ($\mu g/m^3$) of $PM_{2.5}$ and 20 $\mu g/m^3$ of PM_{10} in response to increased evidence of health effects at very low concentrations of PM.[2]

Air Pollution Sources

Four sources of PM monitoring data are utilized here to estimate annual average ambient air concentrations of $PM_{2.5}$ in Karachi. Mansha and others (2011) report $PM_{2.5}$ concentrations at a residential site in Karachi for the period January 2006 to January 2008 (figure 4.1). Ghauri (2008) reports $PM_{2.5}$ and $PM_{1.0}$ concentrations at SUPARCO in Karachi from September 2007 to June 2008 (figure 4.1).

The Sindh Environmental Protection Agency (Sindh EPA) (2010) reports $PM_{2.5}$ concentrations during periods of the year from February 2008 to April 2009 (figure 4.2). Alam and others (2011) report $PM_{2.5}$ concentrations of 185 $\mu g/m^3$ (461 $\mu g/m^3$ PM_{10}) at M. A. Jinnah Road, 151 $\mu g/m^3$ (270 $\mu g/m^3$ PM_{10}) at SUPARCO, and 60 $\mu g/m^3$ (88 $\mu g/m^3$ PM_{10}) at the Sea View site for April–May 2010.

Figure 4.1 PM Ambient Concentrations in Karachi, SUPARCO, 2007–08

μg/m³

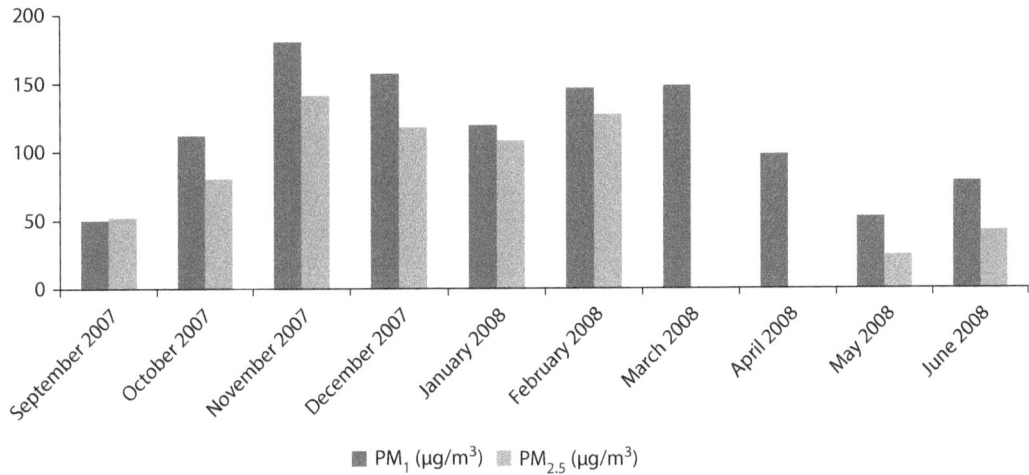

■ PM₁ (μg/m³) ▨ PM₂.₅ (μg/m³)

Source: Ghauri 2008.
Note: PM₁ = particulate matter of less than 1 micron, PM₂.₅ = particulate matter of less than 2.5 microns.

Figure 4.2 PM₂.₅ Ambient Air Concentrations in Karachi, Sindh-EPA, 2008–09

μg/m³

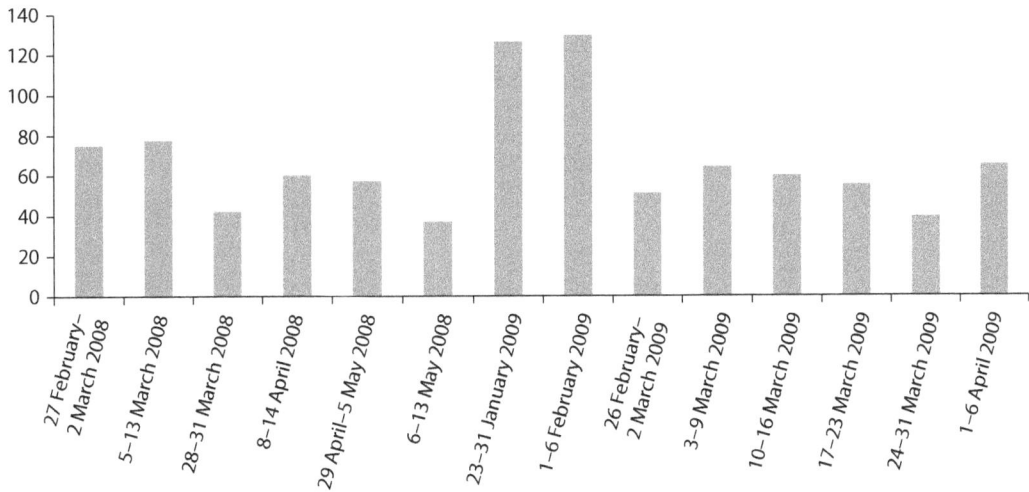

Source: Calculated from daily PM₂.₅ monitoring data reported by Sindh EPA 2010.
Note: PM₂.₅ = particulate matter of less than 2.5 microns.

Overall, the four sources report $PM_{2.5}$ concentrations in the range of 120–180 μg/m³ during the period of November to February, substantially lower concentrations during other months of the year, and 30–50 μg/m³ during June–August. This suggests that annual average $PM_{2.5}$ concentrations in Karachi are about 88 μg/m³ (table 4.1). Annual PM_{10} concentrations may thus be about 183 μg/m³.[3]

Table 4.1 Estimate of Annual Average PM$_{2.5}$ Ambient Air Concentrations in Karachi, 2006–09

	Average PM$_{2.5}$ concentrations (μg/m³)	Number of months
November–February	140	4
March–May	80	3
June–August	40	3
September–October	70	2
Annual average	88	12

Sources: Sánchez-Triana and others 2014. Estimates based on Alam and others 2011, Ghauri 2008, Mansha and others 2011, and Sindh EPA 2010.
Note: PM$_{2.5}$ = particulate matter of less than 2.5 microns.

Map 4.1 Towns and Cantonments of Karachi

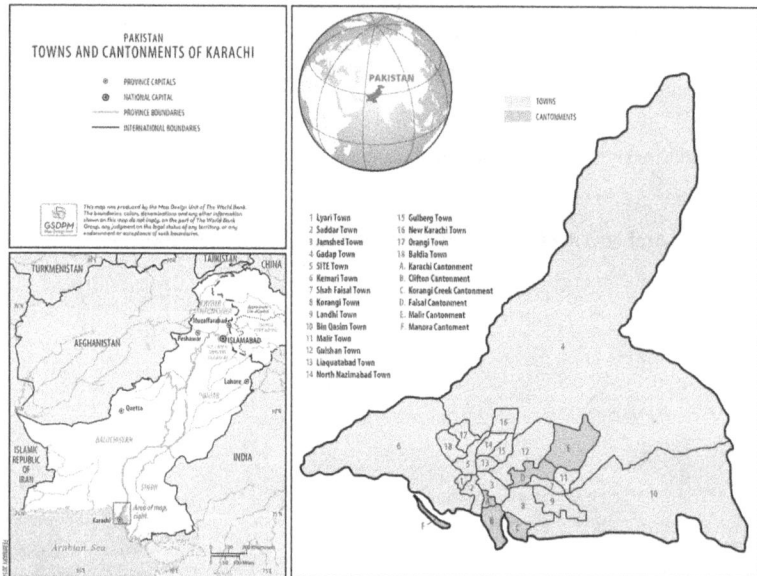

Karachi consists administratively of 18 towns and several cantonments (map 4.1). The total area of Karachi is 3,600 km² with an estimated population of 15.1 million in 2005 (CDGK 2007).[4] About 12.8 million of the population lived in 15 of the towns with an area of 365 km² and an average population density of 35,000 people per km² (table 4.2).[5] This makes Karachi one of the megacities with the highest population density in the world. A very large number of people are therefore exposed to every ton of air pollution emitted in the city.

Wind conditions, rainfall patterns, and other climatic factors influence the contribution to ambient PM concentrations of emissions originating from a specific location in the city. The wind in Karachi is predominantly from the southwest (SW) and the west-southwest (WSW) from March to October, and east-northeast (ENE) and north-northeast (NNE) from November to February. Wind speed is generally highest during May to August and lowest during October

Table 4.2 Population of Karachi by Town and Area, 2005

No.[a]	Town	Population (000)	Area (km²)	Population (000)/km²
1	Lyari	923	8.0	115.4
2	Saddar	936	24.1	38.7
3	Jamshed	1,114	23.4	47.6
5	SITE	710	25.4	27.9
7	Shah Faisal	510	11.7	43.4
8	Korangi	830	41.5	20.0
9	Landhi	1,012	39.1	25.9
11	Malir	605	17.8	34.0
12	Gulshan-e-Iqbal	949	53.7	17.7
13	Liaquatabad	986	10.9	90.7
14	North Nazimabad	753	16.7	45.1
15	Gulberg	689	13.8	49.8
16	New Karachi	1,039	20.5	50.8
17	Orangi	1,099	23.5	46.8
18	Baldia	617	29.2	21.1
	Subtotal (high density towns)	**12,771**	**359**	**35.5**
4	Gadap[b]	440	1,439.9	0.3
6	Kemari	584	429.8	1.4
10	Bin Qasim	481	558.3	0.9
	Cantonment	465	126.8	3.7
	DHA	380	38.3	9.9
	Subtotal (low density towns)	**2,349**	**2,593**	**0.9**

Source: Produced from City District Government Karachi (CDGK 2007).
Note: The total does not exactly correspond to the sum of the items because of rounding in individual items.
DHA = Defense Housing Authority.
a. Number corresponds to number on the map above.
b. Excluding Kirther National Park Area.

to February. Rain falls predominantly during July to September (figure 4.3). These wind and rainfall patterns are likely important factors in explaining why $PM_{2.5}$ ambient concentrations in Karachi are three to four times higher during November to February than during June to August.

An estimated 4 million people lived in 12 other cities with a population of more than 100,000 in 2009 (table 4.3). The estimate reflects an annual average population growth of 3% since the population census in 1998. The estimate is likely an understatement of the entire urban areas of the cities. Monitoring data of PM_{10} or $PM_{2.5}$ are not available for these cities, and thus, World Bank modeling estimates of annual average PM_{10} concentrations were applied (table 4.2).[6] Based on Alam and others (2011) in Karachi, $PM_{2.5}$ is assumed to be 48% of PM_{10} in these cities.

Mortality from PM

The PM concentrations reported above are estimated to be the cause of 20% of acute lower respiratory infections (ALRI) mortality among children under

Figure 4.3 Monthly Average Wind Speed and Rainfall in Karachi

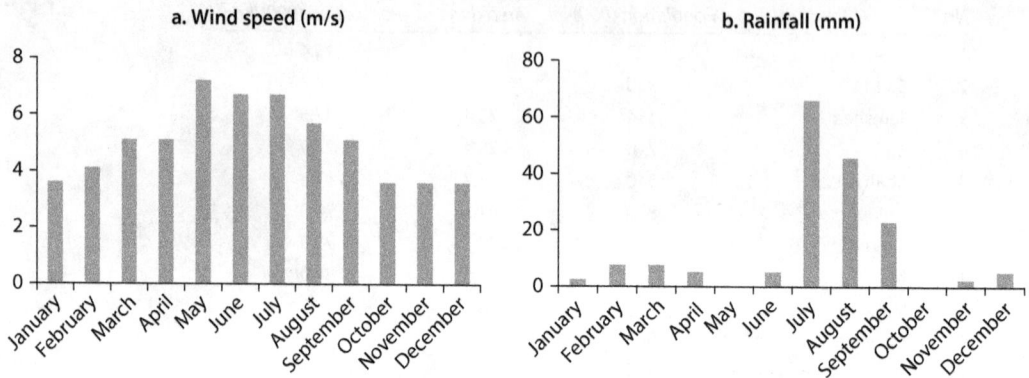

a. Wind speed (m/s)   b. Rainfall (mm)

Sources: www.weather.com; www.windfinder.com.

Table 4.3 Populations and Estimates of Annual Average PM Concentrations in Cities in Sindh

Cities	PM_{10} $(\mu g/m^3)$	$PM_{2.5}$ $(\mu g/m^3)$	Population (millions) 2009
Hyderabad	170	82	1.54
Sukkur	125	60	0.48
Larkana	124	60	0.44
Mirpur Kas	136	65	0.24
Nawabshah	123	59	0.26
Jacobabad	125	60	0.19
Shikarpur	115	55	0.17
Tando Adam	115	55	0.14
Khaipur	114	55	0.14
Dadu, Tando Allah Yar, Khandh Kot[a]	115	55	0.38
Total	n.a.	n.a.	3.98

Sources: PM_{10} concentrations are from World Bank 2006. $PM_{2.5}$ is 48% of PM_{10} based on Alam and others 2011 in Karachi. Population is for 2009, based on estimates reported by World Gazetteer 2011.
Note: $PM_{2.5}$ = particulate matter of less than 2.5 microns, PM_{10} = particulate matter of less than 10 microns, n.a. = not applicable.
a. The World Bank does not report PM_{10} for these cities; PM_{10} concentrations are therefore assumed to be the same as the model estimates for Shikarpur, Tando Adam, and Khajpur.

five years of age, and 24% of cardiopulmonary mortality and 41% of lung cancer mortality among adults 30 years or older in the cities with a population greater than 100,000.[7] This means that over 9,000 individuals die prematurely each year in these cities from PM outdoor air pollution (table 4.4). Nearly 80% of the deaths are in Karachi, and close to 10% are in Hyderabad. About 12% of the deaths are among children under five years of age and 88% are among adults.

Years of life lost (YLL) due to mortality from PM are estimated at nearly 97 thousand annually (table 4.5).[8] About 38% of years lost are among children under five years of age, and 62% are among adults (table 4.5).

Table 4.4 Estimated Annual Mortality from PM Ambient Concentrations in Cities in Sindh, 2009

Mortality	ALRI	Cardiopulmonary	Lung cancer	Total
Population group	Children <5 years of age	Population 30+ years of age	Population 30+ years of age	
Total deaths from PM	1,059	7,752	216	9,027
Karachi (% of total)	81	78	78	79

Source: Sánchez-Triana and others 2014.
Note: ALRI = acute lower respiratory infections; PM = particulate matter.

Table 4.5 Years of Life Lost (YLL) Due to Mortality from PM Ambient Concentrations in Cities in Sindh, 2009

	Population group	YLL per premature death	Total YLL
ALRI mortality	Children <5 years of age	35	37,051
Cardiopulmonary mortality	Population 30+ years of age	7.5	58,137
Lung cancer mortality	Population 30+ years of age	7.5	1,617
Total			96,805

Source: Sánchez-Triana and others 2014.
Note: ALRI = acute lower respiratory infections; PM = particulate matter.

Table 4.6 Estimated Annual Cases of Morbidity from PM Ambient Concentrations in Cities in Sindh, 2009

	Chronic bronchitis	Hospital admissions	Emergency room visits	Restricted activity days	LRI in children	Respiratory symptoms
Karachi	145,185	26,686	523,498	81,838,293	1,353,000	260,459,264
Hyderabad	16,877	2,957	58,003	9,067,520	149,910	28,858,368
Sukkur	4,533	662	12,994	2,031,360	33,584	6,465,024
Larkana	4,138	602	11,808	1,845,888	30,517	5,874,739
Mirpur Kas	2,368	363	7,118	1,112,832	18,398	3,541,709
Nawabshah	2,434	353	6,916	1,081,184	17,875	3,440,986
Jacobabad	1,794	262	5,143	804,080	13,294	2,559,072
Shikarpur	1,534	214	4,202	656,880	10,860	2,090,592
Tando Adam	1,263	176	3,460	540,960	8,943	1,721,664
Khaipur	1,257	175	3,427	535,808	8,858	1,705,267
Dadu Tando, Allah Yar, and Khandh Kot	3,429	479	9,392	1,468,320	24,275	4,673,088
Total	184,814	32,929	645,963	100,983,125	1,669,514	321,389,773
Karachi (% of total)	79	81	81	81	81	81

Source: Sánchez-Triana and others 2014.
Note: The total does not exactly correspond to the sum of the items because of rounding in individual items.

Morbidity from PM

PM concentrations in cities with a population greater than 100,000 in Sindh are also estimated to be the cause of 59% of chronic bronchitis in these cities. This constitutes a total of nearly 185,000 cases, and nearly 33,000 hospital admissions, over 645,000 emergency room visits, over 1.6 million cases of lower respiratory illness (LRI) in children, over 100 million restricted activity days, and over 300 million respiratory symptoms annually (table 4.6).

These health effects represent over 106,000 YLL to disease each year, of which the large majority is among the adult population (table 4.7). Thus, in total,

Table 4.7 Years Lost to Disease (YLD) from PM Outdoor Air Pollution Exposure, 2009

Health endpoint	Population group	YLD/10,000 cases	Total YLD
Chronic bronchitis	Adults 15+ years	2,000	36,963
Hospital admissions	All ages	160	527
Emergency room visits	All ages	45	2,907
Restricted activity days	Adults 15+ years	3	30,295
LRI in children	Children <15 years	70	11,687
Respiratory symptoms	Adults 15+ years	0.75	24,104
Total			106,483

Source: Sánchez-Triana and others 2014.

over 213,000 years of life is lost each year due to PM exposure in the cities in Sindh with more than 100,000 inhabitants.

Lead Exposure

Human exposure to lead (Pb) has many known health effects. Young children are the most vulnerable population group. Impacts of lead on children include neuropsychological impacts (for example, impaired intelligence and increased incidence of mild mental retardation), anemia, and gastrointestinal symptoms (Fewtrell, Kaufmann, and Pruss-Ustun). Health effects of lead have been evidenced by their association with BLL. Lead in blood can originate from exposure to lead in, for instance, air, drinking water, food, dust, soil, paint, cosmetics (for example, surma), utensils, several herbal medicines, children's toys, ornaments and jewelry, and other materials and articles containing lead. Anemia and gastrointestinal symptoms in children generally occur at BLL greater than 60 micrograms of lead per deciliter of blood ($\mu g/dL$), while studies have documented loss of intelligence (IQ [intelligence quotient] points)[9] at BLLs well below 10 $\mu g/dL$. Increased incidence of mild mental retardation from lead is a result of IQ losses that bring a child's IQ below 70 points (Fewtrell, Kaufmann, and Pruss-Ustun 2003). Exposure to lead also has many health effects in adults including increased blood pressure, cardiovascular disease, and reproductive effects (Fewtrell, Kaufmann, and Pruss-Ustun 2003). Since recent data on BLL in adults in Pakistan are scarcer than on BLL in children, and since very few children are likely to have a BLL greater than 60 $\mu g/dL$ today, the focus in this chapter is on IQ losses in children under five years of age.[10]

Blood Lead Levels in Children under Five Years

Average BLLs in children under five years of age in Karachi declined from possibly over 35 $\mu g/dL$ in 1989 to around 20 $\mu g/dL$ in the mid-1990s and further to around 15 $\mu g/dL$ in 2000 (Kadir and others 2008). These declines in BLLs largely correspond to the reduction in the lead content of gasoline in Pakistan from about 1.5 grams per liter (g/L) in 1991 to 0.5 g/L in 1995, 0.4 g/L in 1999, and a phaseout in 2002 (Kadir and others 2008).

A study of children under five years of age in Karachi in 2000 found an average BLL of 12 µg/dL among children in a rural community 50 km from Karachi and 16.5 µg/dL among children in the inner city of Karachi with high traffic density (Rahbar and others 2002). The highest average BLL (21.6 µg/dL) was found among children in an island community outside Karachi (Baba Island). The study identified some of the sources of lead contributing to the elevated BLLs, such as lead in house dust, food, and drinking water.

No published studies have assessed BLLs in representative samples of children less than five years of age in Pakistan after the phaseout of lead in gasoline. A study of children under five years living around automobile and battery repair workshops in Lahore found an average BLL of 10.9 µg/dL (Ahmad and others 2009). A study in Punjab found an average BLL of 9.0 µg/dL among children one to six years of age of smelters/battery recycling plant workers living near the industries in Wah/Gujranwala, and 6.5 µg/dL among children living 30 km from the industrial areas (Khan and others 2010; Khan, Ansari, and Khan 2011).[11] Table 4.8 presents key statistics from these studies.

Recent studies in several large cities in India have found average BLLs of 8–11 µg/dL among children several years after phaseout of lead in gasoline. A study as recent as 2009 found an average BLL of 6.6 µg/dL among rural children in Uttar Pradesh. Children's BLLs seem to be similar in parts of China and rural areas of the Philippines (table 4.9).

In light of the studies in Pakistan, as well as in several other countries in Asia, it may be suspected that average BLLs in children under five years of age in Sindh today may be around 7 µg/dL in urban areas and around 5 µg/dL in rural areas, or less than half of BLLs in 2000. The percentage of children with a BLL greater than 10 µg/dL may be approximately 20% in urban areas and 10% in rural areas. Applying a lognormal distribution of BLL (Fewtrell, Kaufmann, and Pruss-Ustun 2003), an estimated 13% of children under five in Sindh have a BLL greater than 10 µg/dL (table 4.10).[12]

Drinking water in Sindh appears to be a major source of lead exposure. In a study in 2007/08, six samples of household water were taken from piped water taps with water originating from surface water sources and six samples from

Table 4.8 Blood Lead Levels (BLL) in Children in Pakistan

	Khan, Ansari, and Khan 2011	*Khan, Ansari, and Khan 2011*	*Ahmad and others 2009*
Age	children 1–6 years	children 1–6 years	children <4 years
Location	Punjab	Punjab	Lahore
Location	near industry	30 km from industry	near auto and battery repair shops
Sample size	123	123	106
Mean BLL (µg/dL)	9.0	6.5	10.9
Standard deviation	4.5	2.7	5.3
% with BLL >10 µg/dL	31	11	52

Source: Sánchez-Triana and others 2014. Based on the sources indicated in each column.

Cleaning Pakistan's Air • http://dx.doi.org/10.1596/978-1-4648-0235-5

Table 4.9 Blood Lead Levels (BLL) in Children in China, India, and the Philippines

Source	Roy and others 2009	Nichani and others 2006	Ahamed and others 2010	Chaudhary and Sharma 2010	Liu and others 2011	Riddell and others 2007
Age of children	3–7 years	<12 years	3–12 years	3–5 years	3–5 years	<5 years
Location	Chennai, India	Mumbai, India	Lucknow, India	Uttar Pradesh, India	Jintan, South Eastern China	Philippines (rural)
Location	Throughout city	Throughout city	Throughout city	25 rural villages	Urban and rural	11 provinces
Year of study	2005–06	2003	–	2009	2004–05	–
Sample size	756	754	200	100	1,344	2,861
Mean BLL (µg/dL)	11.4	8.4	9.3	6.6	6.4	6.9
% with BLL >10 µg/dL	55	33	37	–	8	21

Source: Sánchez-Triana and others 2014. Based on the sources indicated in each column.
Note: – = Data not provided by the study.

Table 4.10 Estimated Distribution of Children by BLL in Sindh Province

BLL (µg/dL)	Percent of children under five years of age
0–5	39.1
5–10	47.9
10–20	12.5
20–30	0.4
>30	<0.1
Total	100.0

Source: Sánchez-Triana and others 2014.
Note: BLL = blood lead levels.

households using groundwater for drinking in each of 18 districts in Karachi. Lead concentrations in 89% of the 216 samples exceeded the WHO guideline limit of 10 microgram per liter of water (µg/L). The average lead concentration was 77 µg/L in drinking water originating from surface water sources and 146 µg/L in groundwater (table 4.11).[13] In a study of groundwater quality throughout Sindh province, 54% of samples, mainly from pumps and dug wells, contained lead concentrations exceeding the WHO guideline of 10 µg/L. The highest concentration was 111 µg/L (table 4.12).[14]

Applying the BLL model in Sara Group (2008), BLL from lead in drinking water can be estimated for two- to five-year-olds as follows, with a somewhat higher BLL for one-year-olds:

$$BLL = \beta * Pb * w * A \qquad (4.1)$$

where $\beta = 0.36$, *Pb* is lead concentration in water (µg/L), *w* is drinking water consumption (0.5 L/day), and *A* is the absorption rate of Pb in the body (50%). The model predicts that a lead concentration of 10 µg per liter of drinking water (that is, WHO limit value) results in a BLL of about 0.9 µg/dL in these children Lead concentrations of 100 µg/L results in a BLL of about 9 µg/dL (figure 4.4).

Table 4.11 Lead Concentrations in Drinking Water in Karachi, 2007–08

	Surface water sources		Groundwater sources	
Pb concentrations (μg/L)	Mean Pb (μg/L)	Number of districts	Mean Pb (μg/L)	Number of districts
<50	19.6	5	27.5	4
50–100	84.5	7	86.0	3
>100	116.6	6	204.5	11

Source: Sánchez-Triana and others 2014. Produced from data in Ul-Haq and others 2011.

Table 4.12 Lead Concentrations in Groundwater in Sindh Other than Karachi

Pb concentrations (μg/L)	Number of samples	Percent of samples
<10	38	46
10–50	25	31
>50	19	23
Total	82	100

Source: Junejo n.d.

Figure 4.4 Estimated Blood Lead Level (BLL) in Children 2–5 Years Old, in Relation to Pb in Drinking Water

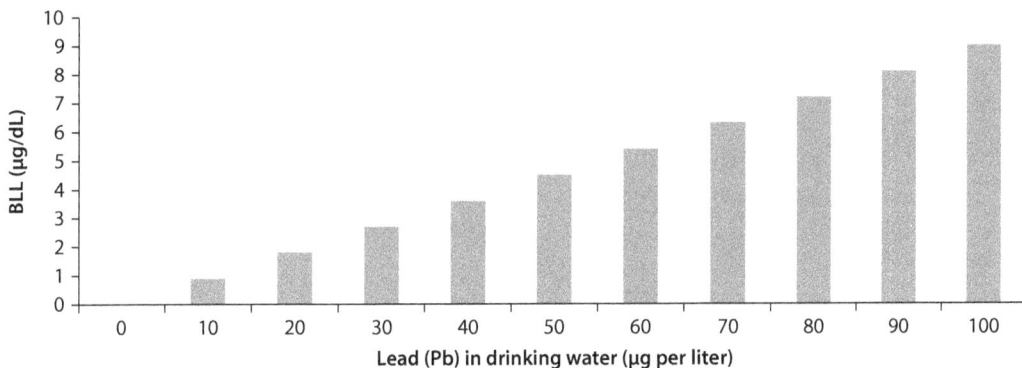

Source: Sánchez-Triana and others 2014. Estimated based on Sara Group 2008.

Based on the model and the lead concentrations measured in drinking water in Sindh, it is estimated that lead in drinking water results in an average BLL of 3–4 μg/dL among children under five years of age in Sindh. This one source of lead exposure may therefore be responsible for over 50% of suspected BLLs among these children (that is, 7 μg/dL in urban and 5 μg/dL in rural children).

Traditional cosmetics (that is, surma) may be another potentially important source of lead exposure. Lead concentration in surma is often very high (> 65%). Rahbar and others (2002) report that 13% of children have surma applied to their eyes at least twice a week in Karachi. Studies in other countries have found that application of surma on children, or use of surma by their mothers increases children's BLL by several μg/dL (Rahbar and others 2002). Children's ornaments and jewelry often also contain lead to which children are exposed. Of 54 tested items of children's jewelry commonly used in India, 20% had a lead content of 28–85% of weight and over 40% had a content exceeding an international safety limit of 0.03% (Toxic Link 2010).

Studies internationally have also found relationships between BLLs in children and their nutritional status. Anemia, or low iron levels, is associated with higher BLLs, and iron supplementation or iron fortification of food given to anemic children has been found to reduce their BLLs. Over 60% of children under five in Pakistan are anemic according to preliminary findings from the National Nutrition Survey 2011.

Neuropsychological Effects in Children under Five Years

Several recent studies have documented neuropsychological effects in terms of intelligence quotient (IQ) losses in children less than five years of age with BLL less than 10 μg/dL (table 4.13).[15] Most of these studies are longitudinal cohort studies that have followed children from infancy to 5–10 years of age, with regular measurement of BLLs and administering of IQ test at age 5–10 years. Lanphear and others (2005) pooled seven international studies in which 17% of children had peak BLLs below 10 μg/dL. One of the pooled studies is Canfield and others (2003). The data from Canfield and others represent over 40% of the sample of children with BLLs below 10 μg/dL in Lanphear and others, and are therefore included in table 4.13. The largest study of children with BLLs below 10 μg/dL is Surkan and others (2007). Impairment of a child's intelligence associated with a BLL of about 10 μg/dL is in the range of 4–7 IQ points according to these studies (table 4.14).

Table 4.13 Recent Studies Assessing the Effect of Blood Lead Level (BLL) <10 μg/dL on Children's IQ Score

	Lanphear and others 2005	Jusko and others 2008	Surkan and others 2007	Canfield and others 2003
Type of study	Longitudinal cohort (7 pooled studies)	Longitudinal cohort	Cross-sectional	Longitudinal cohort
Age of children	<1 to 5–10 years	0.5–6 years	6–10 years	6 months to 5 years
Mean BLL (μg/dL)				
Concurrent	9.7	5.0	2.2	5.8
Lifetime	12.4	7.2		7.4
Peak	18.0	11.4		11.1
Number of children	1,333	174	408	172
Children with BLL <10 μg/dL	244 (peak)	94 (peak)	408 (concurrent)	101 (peak)

Source: Sánchez-Triana and others 2014. Based on the sources indicated in each column.
Note: IQ = intelligence quotient.

Table 4.14 Effect of Blood Lead Level (BLL) <10 µg/dL on Children's IQ Score

	BLL (µg/dL)	IQ-point loss (total)		
		Concurrent BLL	Lifetime BLL	Peak BLL
Lanphear and others 2005	from 2.4 to 10	3.9	–	–
Jusko and others 2008	from <5 to 5–9.9	3.7	4.9	5.6
Surkan and others 2007	from 1–2 to 5–10	6.0	–	–
Canfield and others 2003	from 1 to 10	–	7.4	–

Source: Sánchez-Triana and others 2014. Based on the sources indicated in each row.
Note: – = not measured; IQ = intelligence quotient.

Lanphear and others (2005) found that a log-linear function best describes the relationship between children's IQ and their BLL. Rothenberg and Rothenberg (2005), using the same pooled data as in Lanphear and others, confirmed this. The following log-linear function is therefore applied to estimate IQ losses in children from elevated BLLs:

$$\Delta IQ = \beta \, [\ln (BLL) - \ln(X_0)] \qquad \text{for BLL} \geq X_0 \qquad (4.2a)$$
$$\text{and } \Delta IQ = 0 \qquad \text{for BLL} < X_0 \qquad (4.2b)$$

Lanphear and others report a β = 2.70 (95% CI: 1.66–3.74) for concurrent measurement of BLL, that is, BLL at time of IQ test which is the BLL measurement to which the authors devote most of their analysis. The confidence interval (CI) for β is applied to Sindh to provide a lower and upper estimate of IQ point losses, and with β = 2.70 as a central estimate.

A BLL threshold (X_0) below which there are no impacts on children's IQ has not been identified in the international research literature. Gilbert and Weiss (2006) argue for a BLL action level of 2 µg/dL, and Carlisle and others (2009) for a benchmark of 1.0 µg/dL based on the recent research evidence. With a lower threshold value of X_0 = 2.0 µg/dL, a child five to seven years of age with a concurrent BLL of 10 µg/dL has lost 4.4 IQ points (CI: 2.7–6.0 points) due to lead exposure during the first five years of life. A child with BLL of 20 µg/dL has lost 6.2 IQ points (CI: 3.8–8.6 points). A threshold of X_0 = 1.0 µg/dL implies a loss of an additional 1.8 IQ points (figure 4.5).

The BLLs reported in table 4.7 best correspond to the measurement of lifetime BLLs in longitudinal cohort studies such as Lanphear and others. Concurrent BLLs in Lanphear and others are about 75% of lifetime BLLs. Estimated BLLs among children in Sindh in table 4.9 are therefore multiplied by a factor of 0.75 to arrive at concurrent BLLs applied in equations (3.2a, 3.2b). A lower concurrent threshold value of X_0 = 2.0 µg/dL is applied to Sindh. Total annual losses of IQ points among children less than five years of age are thus estimated at 1.2–2.7 million with a midpoint estimate of 2.0 million (table 4.15).[16] If the threshold value is 1.0–1.5 µg/dL instead of 2.0 µg/dL, the midpoint estimate of annual losses of IQ points is 2.7–3.7 million.

Figure 4.5 Loss of IQ Points in Early Childhood in Relation to Lower Threshold Levels (X_0) of Blood Lead Levels (BLL)

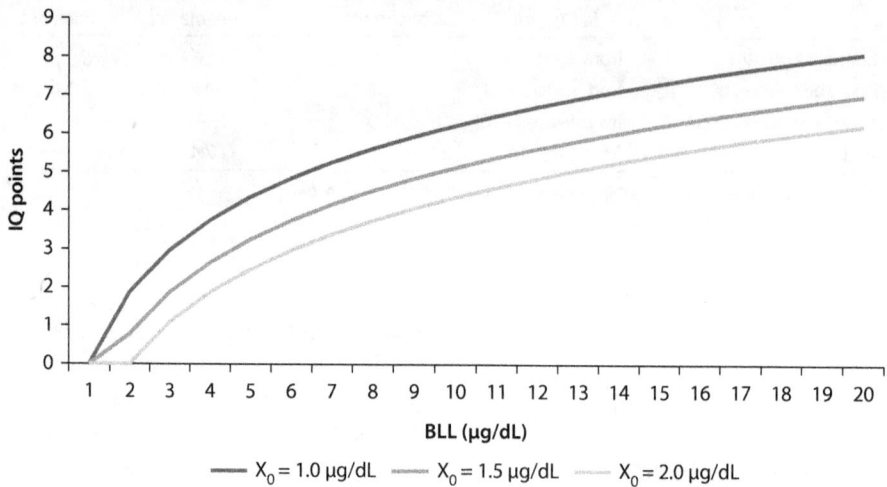

Source: Sánchez-Triana and others 2014. Based on the central estimate of intelligence quotient (IQ) effects of Pb in blood in Lanphear and others 2005.

Table 4.15 Estimated Annual Losses of IQ Points among Children Younger than 5 Years in Sindh, 2009

BLL (µg/dL)	Low (β = 1.66)	Midpoint (β = 2.7)	High (β = 3.74)
<10	1,021,902	1,662,130	2,302,357
10–20	196,705	319,942	443,179
>20	1,702	2,768	3,835
Total	1,220,309	1,984,840	2,749,371

Source: Sánchez-Triana and others 2014.
Note: BLL = blood lead levels; IQ = intelligence quotient.

Social Cost of Health Effects

The health effects estimated in this chapter impose a substantial cost to individuals, households, and society. Income and contributions to household activities are lost from premature mortality, illness, and neuropsychological impairments (IQ losses). Illness also involves cost of medical treatment. These costs can be quantified in monetary terms by means of valuation techniques used in economics. The cost of premature mortality is commonly estimated by using the human capital value (HCV) or a value of statistical life (VSL). The HCV is the present value of lost future income from time of death. The VSL is based on individuals' willingness to pay (WTP) for a reduction in the risk of death. The HCV can also be applied to estimate the cost of IQ losses.

Two valuation techniques are commonly used to estimate the cost of illness (COI). The COI approach includes cost of medical treatment and value of income and time lost to illness. The second approach equates COI to individuals' WTP for avoiding an episode of illness. Studies in many countries have found that individuals' WTP to avoid an episode of an acute illness is generally much higher than the cost of treatment and value of income and time losses (Alberini and Krupnick 2000; Cropper and Oates 1992; Dickie and Gerking 2002; Wilson 2003). A conservative estimate would be that WTP is twice as high as COI.

This study of Sindh province presents the cost of health effects as a range. The lower bound of the range reflects the use of HCV for child and adult mortality and COI for morbidity. The upper bound reflects the use of HCV for child mortality, VSL for adult mortality, and WTP for morbidity (table 4.16). The HCV for adults is substantially lower than the VSL, but nearly the same for children. In the absence of WTP studies in Sindh, WTP is assumed to be twice the COI. The following subsections present estimates of COI.

The cost of health effects of outdoor PM air pollution in cities in Sindh with a population greater than 100,000 is estimated at PRs 30–75 billion per year in 2009, with a midpoint estimate of PRs 53 billion (table 4.17). This cost is equivalent to about 0.8–2.0% of Sindh's gross domestic product (GDP) in 2009, with a midpoint estimate of 1.4% of GDP. Nearly 80% of this cost is from PM pollution in Karachi.

Table 4.16 Economic Cost of a Premature Death Applied to Sindh
Million rupees at 2009 prices

	HCV	VSL	Applied to
Children 0–4 years	4.61	–	OAP, RTA, Noise, WSH, HAP
Children 0–15 years	5.71	–	RTA
Adults (10 years loss of life)	1.27	5.02	OAP, Noise, WSH, HAP
Adults (30 years loss of life)	3.48	5.02	RTA

Source: Sánchez-Triana and others 2014.
Note: HAP = household air pollution, HCV = human capital value, OAP = outdoor air pollution, RTA = road traffic accidents, VSL = value of statistical life, WSH = water, sanitation, and hygiene.

Table 4.17 Estimated Annual Cost of Health Effects of Outdoor PM Air Pollution in Sindh, 2009
Billion PRs

	Low	Midpoint	High
ALRI mortality (children <5 years)	4.9	4.9	4.9
Cardiopulmonary mortality (adults)	9.9	24.4	38.9
Lung cancer mortality (adults)	0.3	0.7	1.1
Chronic bronchitis	2.2	3.3	4.5
Hospital admissions	0.4	0.6	0.8
Emergency room visits	1.1	1.7	2.2
Restricted activity days	4.4	6.7	8.9
Lower respiratory illness in children	3.4	5.1	6.7
Respiratory symptoms	3.5	5.3	7.1
Annual cost (PRs Billion)	30.1	52.7	75.1
Annual cost (% of GDP)	0.81	1.42	2.03

Source: Sánchez-Triana and others 2014.
Note: PM = particulate matter. In cities with population over 100,000.

Mortality accounts for 50–60% and morbidity for 40–50% of estimated cost (table 4.18). The large range in cost of mortality reflects the use of HCV for adults in the "low" estimate of cost and VSL for adults in the "high" estimate. The range in cost of morbidity reflects the use of COI in the "low" estimate of cost and WTP in the "high" estimate, where WTP is twice the COI estimate.

Table 4.19 presents the estimated cost per case of illness using the COI approach. Annual cost of a case of chronic bronchitis (CB) assumes that, annually, 1.5% of individuals with CB have a 10-day hospitalization, 15% have an emergency visit to the doctor, and everyone with CB has one visit to the doctor per year. Time losses are estimated at 2.6 days per person per year. In addition, it is assumed that CB causes a 10% reduction in work productivity. A hospital admission (HAD) is assumed to have a duration of 6 days and result in 10 days of time losses. An emergency room visit (ERV) is assumed associated with 2 days of time losses. LRI in children is assumed to result in a time loss (caretaking by an adult) of 2 hours per day for 15 days. In addition, restricted activity days (RAD) and respiratory symptoms (RS) are assumed to result in a time loss of 1 hour and

Table 4.18 Cost of Mortality and Morbidity from Outdoor PM Air Pollution in Sindh, 2009

	Low	Midpoint	High
Billion PRs			
Mortality	15.0	30.0	44.9
Morbidity	15.1	22.7	30.2
Mortality and morbidity (%)			
Mortality	50	57	60
Morbidity	50	43	40

Source: Sánchez-Triana and others 2014.
Note: PM = particulate matter. In cities with population over 100,000.

Table 4.19 Estimated Annual Cost of Illness (PRs/case), 2009

	Units	Unit cost (PRs)	CB	HAD	ERV	LRI	RAD	RS
Cost of illness								
Hospitalization	per day	1,500	225	9,000	n.a.	n.a.	n.a.	n.a.
Doctor visits	per visit	700	700	n.a.	n.a.	700	n.a.	n.a.
Emergency visits	per visit	1,000	150	n.a.	1,000	n.a.	n.a.	n.a.
Time losses (50% of urban wage rates)	per day	352	900	3,523	705	1,321	44	11
Reduced work productivity[a]	per year	10,076	10,076	n.a.	n.a.	n.a.	n.a.	n.a.
Cost of illness per case (PRs)	n.a.	n.a.	12,051	12,523	1,705	2,021	44	11

Source: Sánchez-Triana and others 2014.
Note: CB = chronic bronchitis; ERV = emergency room visit; HAD = hospital admission; LRI = lower respiratory illness; n.a. = not applicable; RAD = restricted activity days, RS = respiratory symptoms.
a. Calculated as 10% of annual income, with annual income adjusted for non-working adult population (50%).

0.25 hours, respectively, per case. Time losses are valued at 50% of an average urban wage rate of PRs 705 per day. Urban wage rate or income is 1.5 times higher than the average wage rate or income in Sindh according to FBS (2009).

Cost of Lead Exposure

An individual's income is one factor associated with the individual's IQ score. This has long been empirically established, for instance, by Schwartz (1994) and Salkever (1995). These two studies find that a decline of one IQ point is associated with a 1.3–2.0% decline in lifetime income.[17] Studies of the cost of lead (Pb) exposure, or of the benefit of lowering BLL in children, have applied the findings by Schwartz and Salkever in France and the United States, as well as other countries (Gould 2009; Grosse and others 2002; Muennig 2009; Pichery and others 2011).

The present value of future lifetime income of a child under five years is estimated at PRs 4.8 million. This value is estimated based on a real annual future income growth of 2%, assuming that real income in the long run grows at a rate close to the growth rate of GDP per capita. GDP per capita growth in Pakistan was 2% per year from 1990 to 2009, and 2.3% per year from 1970 to 2009 (World Bank 2011).

The cost of a lost IQ point is estimated at PRs 47,000. This is estimated as the product of income loss per lost IQ point (midpoint estimate in Schwartz (1994) and Salkever (1995)) and the percentage of children who may be expected to participate in the labor force. The expected labor force participation is assumed to be the same as the current rate of participation. With an estimated annual loss of 1.2–2.7 million IQ points among children under five in Sindh, the estimated annual cost is PRs 58–130 billion, with a midpoint estimate of PRs 94 billion. This is equivalent to 1.6–3.5% of Sindh's estimated GDP in 2009, with a midpoint estimate of 2.5% of GDP (table 4.20).

The potential benefits of interventions to reduce health effects from the environmental risks assessed in this chapter are substantial, but should be compared to the cost of such interventions. This chapter presents a cost-benefit analysis (CBA) of select interventions for outdoor air pollution (OAP) in Karachi.

Table 4.20 Estimated Annual Cost of IQ Losses among Children Younger than 5 Years in Sindh, 2009

	Low	Midpoint	High
Present value of future lifetime income (15–64 years) (PRs)	4,824,922	4,824,922	4,824,922
Lifetime income loss per IQ point lost (% of lifetime income)	1.66	1.66	1.66
Labor force participation rate (15–64 years)	59.05%	59.05%	59.05%
Cost per lost IQ point (PRs)	47,372	47,372	47,372
IQ points lost per year	1,220,309	1,984,840	2,749,371
Total cost (PRs million)	57,808	94,025	130,243
Cost % of GDP, 2009	1.56	2.54	3.52

Source: Sánchez-Triana and others 2014.
Note: IQ = intelligence quotient.

The results are expressed as benefit-cost ratios (BCRs). A BCR is the present value of benefits divided by the present value of costs of an intervention evaluated over the useful life of the intervention, that is, over the time period or number of years that the intervention is expected to provide benefits.

CBA is increasingly used in many countries for assessing the merits of potential environmental interventions. As such, a CBA can serve as an instrument to establish priorities and guide allocation of public and private resources.

Benefits and costs of environmental interventions are often difficult and time-consuming to quantify comprehensively and accurately. Health improvements are the main benefits of improved outdoor air quality in urban areas, but reduced building damage, improved crop yields, forest and freshwater quality, and visibility are often additional benefits. Some interventions aimed at improving urban air quality may also reduce greenhouse gas (GHG) emissions. Costs of interventions often include public and private costs. Public costs may be costs of programs promoting behavioral change, while private costs may be increased vehicle fuel cost, higher cost of vehicles, or industrial pollution abatement. A CBA should include both public and private costs to reflect the social or economic viability of interventions.

An increasing number of CBA studies from developing countries shed light on potential benefits and costs of various interventions to improve urban air quality. These studies include Blumberg and others (2006) from China; Larsen (2005) for Bogotá, Colombia; Stevens, Wilson, and Hammitt (2005) from Mexico City, Mexico; ECON (2006) from Lima, Peru; Larsen (2007) from Dakar, Senegal; Larsen (2009) from the Philippines; and Larsen (2011) from the Middle East and the North Africa region. The Larsen, Hutton, and Khanna (2009) report reviews most of these studies. The focus of these studies is on PM, since it is considered the pollutant with the largest health effects in most urban areas. Interventions evaluated in these studies are control of emissions from motorized transport. There are also CBA studies in developing countries evaluating the control of emissions from industry and power plants.

Interventions to Improve Air Quality

Nearly 80% of estimated health effects from PM in cities with population greater than 100,000 inhabitants in Sindh occur in Karachi (see table 4.4 above). The assessment of benefits and costs of interventions to improve PM air quality therefore focuses on Karachi.

Identifiable source-specific PM emissions in Karachi are estimated at 20–28 thousand tons of PM_{10} per year, of which 16–22 thousand tons are $PM_{2.5}$. About 70–80% of these emissions are estimated to end up in the urban air shed of Karachi, that is, 15–19 thousand tons of PM_{10} of which 13–16 thousand tons are $PM_{2.5}$. The remaining emissions drift away from the city, depending on such factors as location of emission source and wind conditions. With the addition of secondary particulates (sulfates and nitrates)[18] and PM from area-wide sources, total annual PM_{10} that causes the PM ambient concentrations in the urban air shed of Karachi is estimated at 43–51 thousand tons, of which 21–25 thousand tons are $PM_{2.5}$ (see table 4.21). The secondary particulates form in the atmosphere from sulfur dioxide and nitrogen oxide emissions originating from fuel combustion and industrial processes. The area-wide sources of PM include natural dust carried by the wind from outside the city, resuspended road dust, dust from the construction sector, agricultural residue burning, and salt particles from the sea.

Road vehicles and industry followed by solid waste burning are the largest sources of identifiable source-specific PM emissions (table 4.22). These sources are responsible for an estimated 29–36% of PM_{10} and 51–62% of $PM_{2.5}$. About two-thirds of PM from industry appears to be from iron and steel and metal smelters (see Mansha and others 2011). As much as 56–58% of PM_{10} is from area-wide sources, but these sources are less than one-quarter of $PM_{2.5}$. The difference stems from the fact that most PM from area-wide sources consists of larger particulates. Improved street cleaning, containment of dust from the

Table 4.21 Estimated PM Emissions in Karachi (thousand tons per year), 2009

	Urban air shed PM_{10}		Urban air shed $PM_{2.5}$	
	Low	High	Low	High
Source-specific emissions	15.2	19.4	13.1	15.8
Secondary particulates (sulfates, nitrates)	2.7	3.2	2.6	3.0
PM from area-wide sources[a]	25.1	28.5	5.0	5.7
Total	43.0	51.1	20.7	24.6

Source: Sánchez-Triana and others 2014.
Note: PM = particulate matter; $PM_{2.5}$ = particulate matter of less than 2.5 microns; PM_{10} = particulate matter of less than 10 microns.
a. Includes natural dust, sea particles, construction dust, resuspended road dust, and agricultural residue burning.

Table 4.22 Estimated Source Contribution to Ambient PM Concentrations in Karachi, 2009
Percent

	PM_{10}	$PM_{2.5}$
Road vehicles	12–14	24–28
Industry	13–15	19–20
Solid waste burning	4–7	8–14
Domestic (wood/biomass)	2.3–2.7	4–5
Domestic/public/commercial (oil, gas consumption)	0.9–1.0	1.5–1.7
Power plants	0.5–0.6	0.8–0.9
Secondary particulates (sulfates, nitrates)	6–7	12–13
PM from area wide sources[a]	56–58	23–24
Total	100	100

Source: Sánchez-Triana and others 2014.
Note: PM = particulate matter, $PM_{2.5}$ = particulate matter of less than 2.5 microns; PM_{10} = particulate matter of less than 10 microns.
a. Includes natural dust, sea particles, construction dust, resuspended road dust, and agricultural residue burning.

Table 4.23 Estimated Annual PM$_{10}$ Emissions in Karachi from Road Vehicles, 2009

	Number of vehicles	PM$_{10}$ (tons/year)	PM$_{10}$ (% of total)
Rickshaws (two-stroke)	50,000	700	11
Motorbikes (two-stroke)	905,000	1,425	23
Small trucks (diesel) (vans and pickups)	96,000	1,296	21
Minibuses (diesel)	11,500	644	11
Buses, large (diesel)	3,000	185	3
Heavy-duty trucks (diesel)	28,000	1,680	27
Other vehicles	917,500	203	3
Total	2,011,000	6,133	100

Source: Sánchez-Triana and others 2014.
Note: PM$_{10}$ = particulate matter of less than 10 microns.

construction sector, and restrictions on agricultural residue burning can potentially control a share of PM from area-wide sources.

Annual emissions of PM$_{10}$ in Karachi from road vehicles are estimated at over
6,100 tons in 2009 (table 4.23). Two-stroke motorcycles and rickshaws contribute about a third of these emissions, diesel-fueled minibuses and small diesel
trucks (vans, pickups) another third, and heavy-duty trucks over a quarter. Large
buses contribute 3%, and all other vehicles about 3%. These emission estimates
are very rough orders of magnitude; these estimates depend on the accuracy of
figures regarding the number of vehicles and applied emission factors.

The health benefits per ton of PM$_{10}$ emission reductions are on average
estimated at PRs 0.5–1.2 million (US$6.2–15.5 thousand).[19] Benefits of PM
emission reductions are highest for emissions from road vehicles; oil and gas
consumption in the domestic, public, and commercial sectors; and domestic use
of wood/biomass. This is because almost all of the emissions from these sources
end up in the urban air shed and the emissions have a very high PM$_{2.5}$ fraction,
thus high mortality reduction benefits. The source-specific benefits are applied
when assessing the benefits and costs of a specific intervention. It should be
noted, however, that reducing PM$_{10}$ emissions from a source with low benefits
per ton of emissions might still have high benefits relative to costs if the cost of
intervention is relatively low.

Implementation of well-targeted interventions could possibly improve ambient PM$_{2.5}$ air quality in Karachi by 40–50% over the next 10–15 years. Reducing
PM emissions from road vehicles will be an essential component, as road vehicles
are estimated to contribute 24–28% of PM$_{2.5}$ in the city. Addressing PM from
industry, solid waste burning, and area-wide sources is also essential. The focus of
this chapter is on road vehicles.

A strategy and action plan to control PM emissions from road vehicles will
need to address in-use vehicles, and new and imported secondhand vehicles
entering the roads. As a large share of PM emissions from road vehicles is from
diesel vehicles, an essential action is to limit the sulfur in diesel. A first step is to
reduce sulfur in diesel to 500 ppm, and a second step is to reduce it further to
50 ppm. Reducing sulfur in diesel has direct benefits in terms of lowering PM

emissions from all diesel vehicles regardless of their vintage and (lack of) PM control technology, and indirect benefits in terms of allowing retrofitting of in-use vehicles with PM control technology and ensuring that new and imported secondhand vehicles comply with modern emission standards. This is very similar to the arguments for phasing out lead (Pb) in gasoline some years ago. Phasing out lead had direct benefits in terms of mitigating the health effects of lead emissions and indirect benefits in terms of allowing the installation of catalytic converters on gasoline vehicles.

Some countries around the world have first supplied low-sulfur diesel in major urban areas in a transition phase. This allows retrofitting of diesel vehicles such as buses, taxis, and trucks exclusively used in those urban areas. It is, however, not sufficient for new vehicles with PM control technology that may travel and fuel up in parts of the country that supply only higher-sulfur diesel.

Pakistan has the largest number of compressed natural gas (CNG) vehicles in the world. A large share of passenger cars has already been converted to CNG. Promoting conversion of diesel minibuses, diesel vans, and two-stroke rickshaws could potentially yield substantial PM emission reductions. Nearly half of all vehicles in Karachi are motorcycles, of which a vast majority has highly polluting two-stroke engines. Addressing this source of PM pollution is therefore essential.

The interventions assessed in this chapter are

1. Reducing sulfur in diesel and fuel oil;
2. Retrofitting in-use diesel vehicles with PM emission-control technology;
3. Converting diesel-fueled minibuses and vans to CNG;
4. Controlling PM emissions from motorcycles; and
5. Converting three-wheeled vehicles (rickshaws) to CNG.

These interventions would not only reduce PM emissions but low-sulfur fuels would also reduce secondary particulates by reducing sulfur dioxide emissions. Low sulfur diesel will also allow more stringent emission standards on new diesel vehicles. Other potential interventions that should be assessed include curtailing burning of solid waste in the city (with consideration to the informal recycling industry), controlling PM emissions from ferrous metal sources and other industrial sources, improving street cleaning, and controlling dust from the construction sector.

Low-Sulfur Diesel and Fuel Oil

The world is moving rapidly to using diesel with low-sulfur content in transportation, as well as in other sectors. Low-sulfur diesel reduces PM emissions from combustion and allows installation of very effective PM emission-control technology on diesel vehicles. The European Union (EU) restricted the maximum allowable sulfur content to 500 ppm (0.05%) in diesel in 1996 and to 10 ppm in 2009 (table 4.24).

Many developing countries, including Asian countries, have already implemented 500-ppm sulfur diesel standards (for example, India, Malaysia, the

Table 4.24 Maximum Allowable Sulfur Content in Vehicle Diesel Fuel in the European Union

	Year	Maximum sulfur content (ppm)
Euro 1	1994	2,000
Euro 2	1996	500
Euro 3	2000	350
Euro 4	2005	50
Euro 5	2009	10

Source: http://www.dieselnet.com/standards/eu/ld.php.

Table 4.25 PM Emission Reductions from Lowering the Sulfur Content in Diesel

	Vehicle PM emission reductions	Main sources of data
Diesel (500-ppm sulfur)	>20%	ADB 2005
Diesel (50-ppm sulfur)	>33%	Blumberg, Walsh, and Pera 2003; UNEP 2006

Source: Sánchez-Triana and others 2014. Based on the sources indicated in each row.
Note: PM = particulate matter.

Philippines, Thailand, and Vietnam). Some countries have introduced 350-ppm or 50-ppm sulfur diesel in some metropolitan areas (for example, China and India), and some countries are in the process of introducing or have established timetables for 50-ppm sulfur diesel (for example, Malaysia, the Philippines, Thailand, and Vietnam) (UNEP 2011).

Sulfur in diesel is being reduced to 500 ppm in Pakistan, but no confirmed timetable has been established for 50-ppm sulfur diesel. Lowering the sulfur content in diesel from 2,000 ppm to 500 ppm reduces PM emissions from diesel vehicles on average by at least 20%. If the sulfur content is greater than 2,000 ppm, then the emission reductions will be even greater. Lowering the sulfur content from 500 ppm to 50 ppm can reduce PM emissions by at least another one-third (table 4.25).

This chapter estimates that the health benefits of using 500-ppm diesel in road transport amounts to at least US$2.3–3.5 per barrel of diesel for light diesel vehicles and large diesel buses and trucks used primarily within Karachi. Lowering the sulfur content further to 50 ppm would provide additional health benefits of US$3.0–4.6 per barrel. These estimates are based on PM emission reductions of 20% and 33% for diesel with 500-ppm and 50-ppm sulfur content, respectively.

The additional cost of providing 500-ppm and 50-ppm sulfur diesel depends on several factors related to cost of refinery upgrading and market conditions. A study of 145 refineries in 12 Asian countries estimated the cost of refinery upgrading to produce lower-sulfur diesel. Bringing diesel sulfur content from mostly >5,000 ppm to 500 ppm would cost an additional US$1.3–2.4 per barrel. Bringing the sulfur content from 3,000 ppm to 500 ppm would cost US$0.5–1.1 per barrel. In addition, bringing diesel sulfur content from 500 ppm to 50 ppm would cost an additional US$2–3 per barrel (Enstrat 2003). In the wholesale petroleum product market in the United States, in the last two years, the price difference between >500-ppm diesel and 15–500-ppm diesel was on average about US$2.7 per barrel and the difference between 15–500-ppm diesel and <15-ppm diesel was on average US$1.5 per barrel (EIA 2012). An incremental cost range of US$1.5–2.5 per barrel for lowering the sulfur content to 500 ppm and US$2–3 per barrel for lowering sulfur from 500 ppm to 50 ppm is applied in the assessment in this chapter.

Based on the above, estimated health benefits per dollar spent (that is, BCR) on cleaner diesel are in the range of about US$1–1.5 for light-duty diesel vehicles and US$1.5–2.4 for large buses and trucks for both 500-ppm and 50-ppm diesel (figure 4.6).

Fuel oil in Pakistan generally has a sulfur content that averages around 3%, but some fuel oil with 1% sulfur is being imported. Sulfur content greatly influences PM emission rates from combustion of fuel oil. This book estimates that reducing sulfur from 3% to 1% would have health benefits of US$35–47 per ton of fuel oil, depending on the location in the city where the fuel is used. The additional cost of low-sulfur fuel oil in the international markets fluctuates. The additional cost in Europe of 1% versus 3.5% sulfur fuel oil was around US$30 per ton in 2010 and US$40–50 per ton in 2011 (Bloomberg 2010; Reuters 2012). SNC-Lavalin (2011) estimated an incremental cost of US$50 per ton. Thus, use of low-sulfur fuel oil in Karachi should be targeted at users within the city where all PM emissions contribute to PM ambient concentrations and health benefits are highest. There are, however, additional health benefits of low-sulfur fuel oil that this chapter does not assess. These additional benefits include reduced sulfur dioxide emissions and thus lower secondary particulates formation.

The majority of diesel vehicles in the high-income countries of the EU are fitted with EURO 3–5 PM emission-control technology. Most diesel vehicles in Pakistan do not comply with EURO 1 PM emission standards because of high sulfur content in diesel. Now that diesel with 500-ppm sulfur content will be the

Figure 4.6 Benefit-Cost Ratios of 500-ppm Sulfur Diesel (Left) and 50-ppm Sulfur Diesel (Right) in Karachi

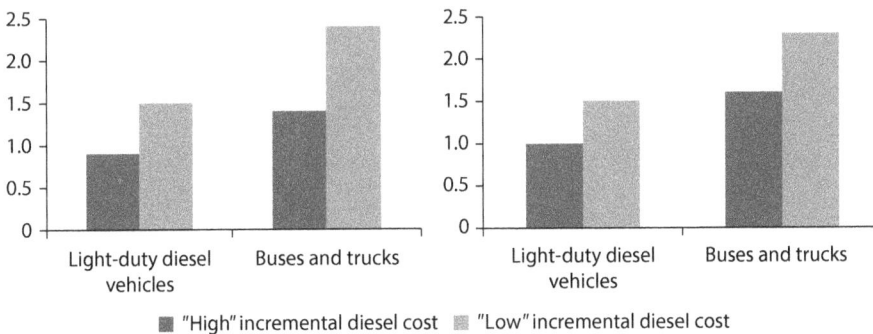

"High" incremental diesel cost   "Low" incremental diesel cost

Source: Sánchez-Triana and others 2014.

norm in Pakistan, new (and secondhand imported) diesel vehicles can comply with EURO 2 emission standards. This will reduce PM emissions quite substantially even compared with EURO 1 standards (figures 4.8 and 4.9).

Reducing sulfur in diesel to 50 ppm would allow implementation of EURO 4 standards. PM emissions per km driven by EURO 1 compliant diesel passenger and light commercial vehicles (LCV) are 4.2–5.6 times higher than PM emissions from EURO 4 compliant vehicles (figure 4.7). For heavy-duty diesel vehicles,

Figure 4.7 European Union Diesel Vehicle Emission Standards for PM
g/km

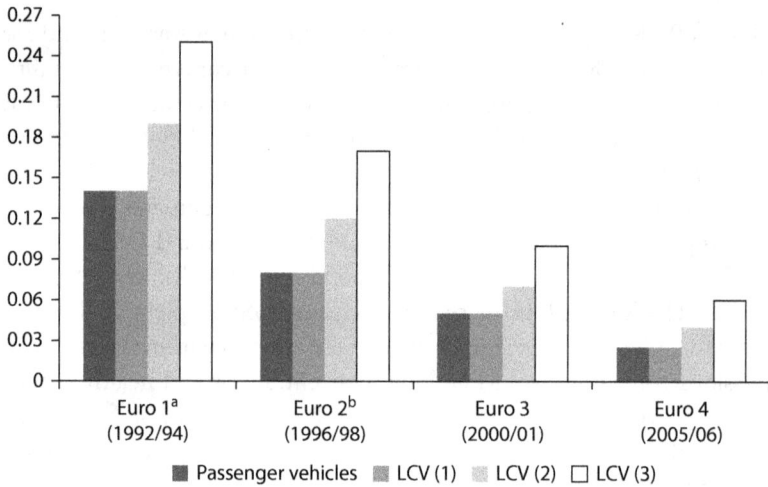

Legend: Passenger vehicles   LCV (1)   LCV (2)   LCV (3)

Source: Adapted from www.dieselnet.com.
Note: PM = particulate matter; IDI = Indirect injection; DI = Direct injection.
a. The earlier year is for passenger vehicles. The later year is for light commercial vehicles (LCV) by weight class 1–3.
b. Applicable for IDI engines. Slightly less stringent limits apply for DI engines.

Figure 4.8 European Union Heavy-Duty Diesel Engines Emission Standards for PM
g/kWh

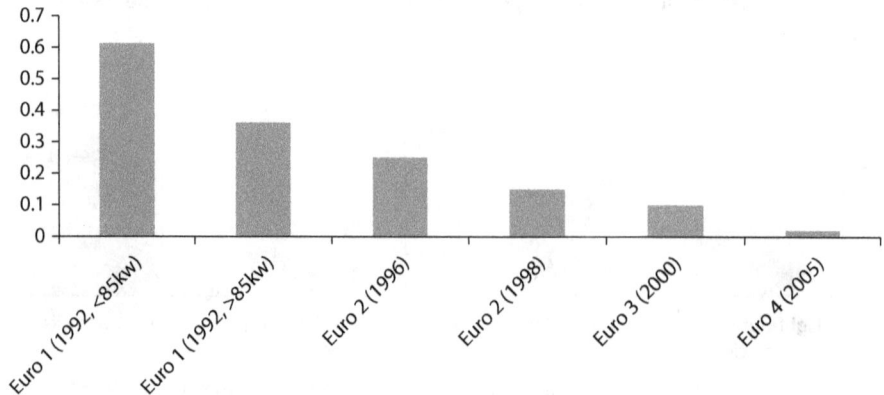

Source: Adapted from www.dieselnet.com.

Figure 4.9 Benefit-Cost Ratios of Retrofitting In-Use Diesel Vehicles with DOC

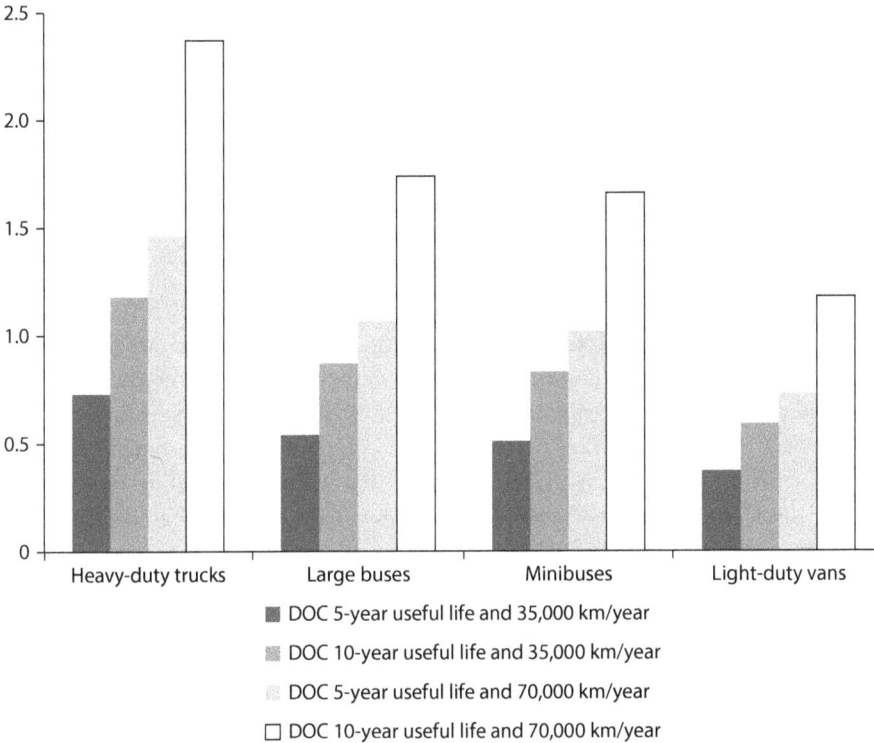

Source: Sánchez-Triana and others 2014.
Note: DOC = diesel oxidation catalyst.

the situation is even worse, since those compliant with EURO 1 emit 18–30 times
more PM per kWh of engine power than EURO 4 compliant vehicles (figure 4.9).

Retrofitting of In-Use Diesel Vehicles

PM control technologies can effectively be installed on in-use diesel vehicles,
such as diesel oxidation catalysts (DOC) and diesel particulate filters (DPFs),
once low-sulfur diesel is available. DOCs require a maximum of 500-ppm sulfur
in diesel and DPFs require a maximum of 50 ppm to function effectively.

DOCs had been installed on over 50 million diesel passenger vehicles and
more than 1.5 million buses and trucks worldwide as of approximately eight
years ago (UNEP 2006). All new on-road diesel vehicles in the United States and
Canada are equipped with a high-efficiency DPF. In addition, all new diesel cars
and vans in the EU were equipped with DPF from 2009 on. Worldwide, over
250,000 heavy-duty vehicles have been retrofitted with DPF (UNEP 2009).
DOCs and DPFs have also been used for retrofitting of buses and trucks in many
countries and locations on a wider scale or in demonstration projects (table 4.26).

Retrofitting EURO 1 or pre-EURO 1 diesel vehicles with a DOC generally
reduces PM emissions by 20–30%, although reductions of as much as 20–50%

Table 4.26 Use of DOC and DPF on Diesel Vehicles for PM Emission Control

Technology	Examples of implementation
DPF on new diesel vehicles	Europe, United States, Canada
DOC and DPF retrofitting of in-use vehicles	Chile; China; Europe; Hong Kong SAR, China; India; Japan; Mexico; Taiwan, China; Thailand; United States

Source: UNEP 2006.

Note: DOC = diesel oxidation catalysts; DPF = diesel particulate filters; PM = particulate matter.

have been reported. A DPF reduces PM emissions by more than 80% (table 4.27). Potential candidates for a DOC, or for a DPF once 50-ppm sulfur diesel becomes available, are high-usage commercial diesel vehicles that are on the roads of Karachi today and primarily used within the city. This book estimates the health benefits of retrofitting per vehicle per year to be in the range of about US$95–568 for a DOC and US$216–1,295 for a DPF depending on type of vehicle and annual usage (table 4.28). These estimates are based on a PM emission reduction of 25% for DOCs and 85% for DPFs.

A DOC costs US$1,000–2,000 and a DPF as much as US$ 6,000–10,000 (MECA 2009; UNEP 2009; USEPA 2012). The expected number of years that the vehicle will continue to be in use and years that the devices will be effective is therefore an important consideration. Estimated health benefits of a DOC only exceed its cost (that is, BCR > 1) for retrofitting of very high usage vehicles (for example, 70,000 km/year) and a relatively long useful life of the DOC (for example, 10 years). If the DOC's useful life is only 5 years, then health benefits only exceed the cost of the DOC for heavy-duty trucks, even when vehicles are used 70,000 km per year (figure 4.9).[20] The estimated health benefits of a DPF are currently lower than its cost for all classes of diesel vehicles, but should be reassessed once 50-ppm sulfur diesel is available in the future.

Conversion of Light-Duty Diesel Vehicles to CNG

Given the relatively high cost of DOCs per kilogram of PM emission reduction, alternative options can be considered for in-use diesel-fueled minibuses and light-duty vans. One such option is conversion to CNG, which almost entirely removes PM emissions. This book estimates the health benefits of CNG conversion to be in the range of about US$455–1,288 per vehicle per year depending on type of vehicle and annual usage (table 4.29).

Conversion of such vehicles to CNG in Pakistan is reported to cost in the range of PRs 150–200 thousand per vehicle, or US$1,900–2,550 at exchange rates in 2009 (Daily Times 2012). Applying a cost of PRs 200 thousand, estimated health benefits per dollar spent on conversion to CNG (that is, BCR) are in the range of US$1–3 for diesel minibuses and US$0.7–2.2 for light-duty diesel vans depending on the length of expected remaining useful life of vehicles and their annual usage (figure 4.10).[21] The BCRs for vans are somewhat lower than for minibuses due to a difference in estimated PM emissions per kilometer of vehicle use.

Table 4.27 PM Emission Reductions from Retrofitting of In-Use Diesel Vehicles

	Vehicle PM emission reductions	Main sources of data
Diesel oxidation catalyst (DOC)	20–30% (20–50%)	UNEP 2009; MECA 2009
Diesel particulate filter (DPF)	>85%	UNEP 2009

Source: Sánchez-Triana and others 2014. Based on the sources indicated in each row.
Note: PM = particulate matter.

Table 4.28 Estimated Health Benefits of Retrofitting Diesel Vehicles
US$/vehicle/year

	Diesel oxidation catalyst (DOC)		Diesel particulate filter (DPF)	
	Vehicle usage		Vehicle usage	
	35,000 km/year	70,000 km/year	35,000 km/year	70,000 km/year
Heavy-duty trucks	284	568	647	1,295
Large buses	208	417	475	949
Minibuses	133	265	302	604
Light-duty vans	95	189	216	432

Source: Sánchez-Triana and others 2014.

Table 4.29 Estimated Health Benefits of Conversion to CNG
US$/vehicle/year

	Vehicle usage	
	35,000 km/year	70,000 km/year
Minibuses	644	1,288
Light-duty vans	455	909

Source: Sánchez-Triana and others 2014.
Note: CNG = compressed natural gas.

Figure 4.10 Benefit-Cost Ratios of Converting In-Use Diesel-Fueled Minibuses and Vans to CNG

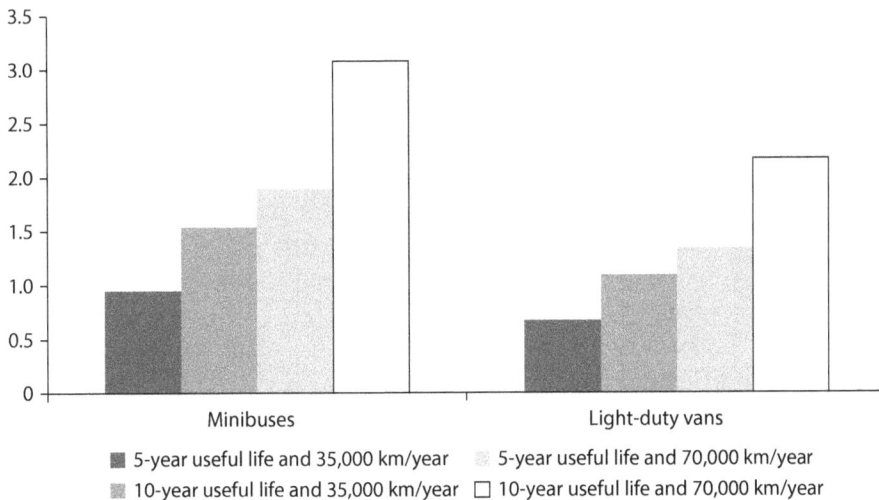

■ 5-year useful life and 35,000 km/year   ▨ 5-year useful life and 70,000 km/year
▨ 10-year useful life and 35,000 km/year   □ 10-year useful life and 70,000 km/year

Source: Sánchez-Triana and others 2014.
Note: CNG = compressed natural gas.

Controlling PM Emissions from Motorcycles and Rickshaws

Two-stroke motorcycles tend to have high PM emissions per ton of gasoline consumption. They also contribute greatly to urban noise. Many countries have limited or banned the use of two-stroke motorcycles. Four-stroke motorcycles emit substantially less PM, are more fuel-efficient, and cause substantially less noise. The additional cost of a four-stoke motorcycle engine is approximately US$50, but varies with engine size (Meszler 2007).[22]

At this cost, and for a two-stroke motorcycle driven 7,000 km per year and emitting 0.25 grams of PM per kilometer, the health benefits of switching to a four-stroke motorcycle is around US$2.9 per dollar of additional engine cost, if emissions reductions are about 50%. With fuel savings the BCR is as high as 4.2.

Two-stroke rickshaws are also a large source of PM emissions and urban noise. Conversion to a four-stroke engine using CNG is an option and is reported to cost around PRs 40–60,000 in Pakistan. At a cost of PRs 60,000 for a rickshaw that is driven 40,000 km per year and emitting 0.35 grams of PM per kilometer, the health benefits are estimated at 1.6 –2.6 times the conversion cost for a rickshaw with a remaining life of 5–10 years.[23]

Conclusions and Recommendations

This chapter provided an assessment of benefits and costs of selected environmental health interventions to control PM air pollution in Karachi. Benefits per rupee spent (that is, BCR) on lowering the sulfur content in diesel to 500 ppm and further to 50 ppm are estimated at PRs 1.1–1.2 for light-duty diesel vehicles and PRs 1.7–1.8 for large buses and trucks used in Karachi.[24] Benefits per rupee spent on lowering the sulfur content in fuel oil to 1% are about the same as cost if the fuel oil is used within Karachi and less than cost if the fuel oil is used in the outskirts of the city (figure 4.11).

The benefits per rupee spent on retrofitting in-use diesel vehicles with a DOC, once 500-ppm diesel is available, are estimated at PRs 1–1.3 for large buses and trucks used within the city but less than cost for minibuses and light-duty vans (figure 4.12). Benefits per rupee spent on converting diesel minibuses and light-duty vans to CNG are, however, estimated at PRs 1.2–1.7 (figure 4.13).

The benefits of converting two-stroke rickshaws to four-stroke CNG are about twice as high as conversion cost, and the benefits of four-stroke motorcycles (instead of two-stroke) are nearly three times higher than the additional cost of a four-stroke engine.

While the interventions discussed above can substantially reduce PM emissions from road vehicles in Karachi, interventions to control PM from other sources must also be considered. These include effectively enforcing the ban on burning of solid waste in the city; moving any existing brick kilns, metal foundries, and scrap smelters out of the city; and considering wind directions and future urban development when deciding on acceptable locations. Additional interventions include improving street cleaning to reduce resuspension of road

Figure 4.11 Benefit-Cost Ratios of Low-Sulfur Fuels in Karachi

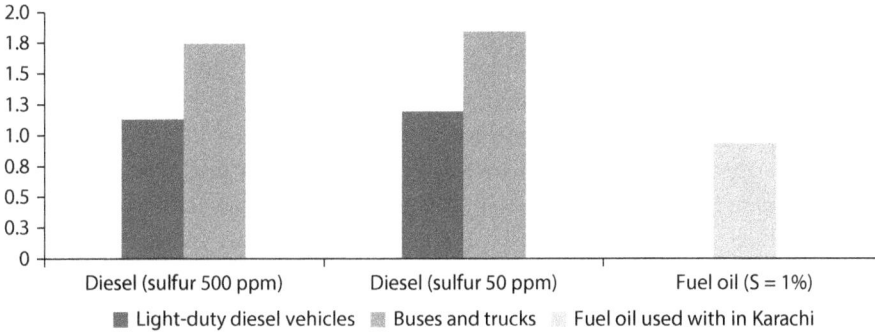

Source: Sánchez-Triana and others 2014.
Note: Midpoint estimate of incremental cost of low-sulfur fuels. S is the sulfur content in fuel oil.

Figure 4.12 Benefit-Cost Ratios of Retrofitting In-Use Diesel Vehicles with DOC in Karachi

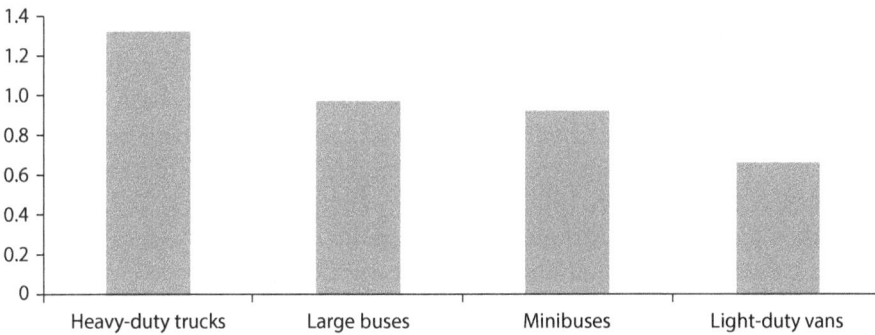

Source: Sánchez-Triana and others 2014.
Note: Seven years of useful life of DOC; 50,000 km/year vehicle usage. DOC = diesel oxidation catalysts.

Figure 4.13 Benefit-Cost Ratios of Conversion of Diesel Vehicles, Rickshaws, and Motorcycles in Karachi

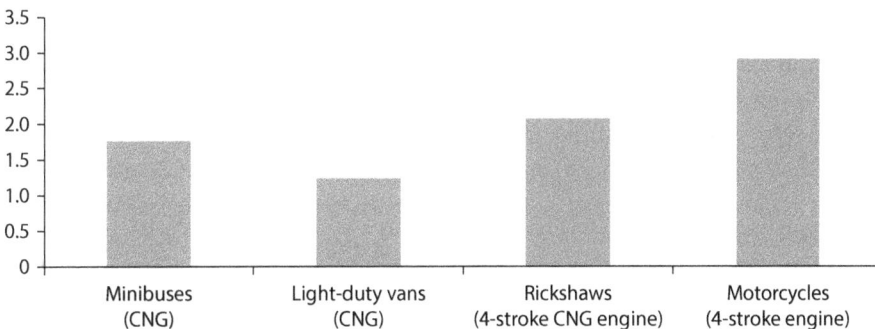

Source: Sánchez-Triana and others 2014.
Note: Seven years of useful life of compressed natural gas (CNG) conversions, and 50,000 km/year usage of minibuses and light-duty vans.

dust and implementing measures to reduce construction dust; and controlling emissions from large point sources (ensuring that existing control equipment is operating properly and installing new where nonexistent).

In the medium term, the GoP might consider adopting demand-side management measures to reduce the country's motorization trend. Key among these measures would be the development of mass transportation in Pakistan's main cities. Experiences from countries such as Brazil, Colombia, and Mexico demonstrate the benefits of relatively new public transport systems, such as Bus Rapid Transit, which can use bus-based technologies to transport increasingly larger volumes of customers at moderately high speeds even in very congested urban areas. While still substantial, the investments needed to develop and operate these systems are significantly lower than the investments needed for traditional mass transport systems, such as underground metros. In addition, these systems have been able to demonstrate their contributions to reduce congestion and pollution, and some of them have even received international funding for their role in reducing GHG emissions from mobile sources.

Additional policies that are worth assessing in the medium term include traffic control, restricted circulation of private cars during high pollution episodes, urban planning and land use, establishment of high-occupancy vehicle lanes, measures to improve traffic flow such as 'green wave' coordination of traffic signals, and improvement of infrastructure, such as paving of roads and regular sweeping. City governments would be responsible for most of these measures, and thus capacity-building efforts would be needed targeting key cities to strengthen their capacity for urban planning and transportation management. In addition, the GoP might consider exploring options to promote the adoption of modern technology spark-ignition engines.

In summary, this chapter presents the following conclusions and guidance:

- The move to 500-ppm sulfur in diesel in Pakistan has larger health benefits than the additional fuel cost.
- Establish a timetable for moving to 50-ppm sulfur in diesel, as many other Asian countries have done.
- Continue to convert diesel minibuses and city delivery vans to CNG and install DOC on existing large buses and trucks used in the city.
- Continue introducing new CNG full-size buses, as DPFs cannot be used with 500-ppm sulfur in diesel.
- Convert existing two-stroke rickshaws to four-stroke CNG engines.
- If CNG supply is a constraint, then give preference for CNG conversion to diesel minibuses, commercial light-duty diesel vans, and rickshaws, rather than gasoline cars.
- Implement and enforce a ban on new two-stroke motorcycles and rickshaws and explore options to control PM emissions from in-use two-stroke motorcycles.
- Introduce low-sulfur fuel oil (1% sulfur) to major users located in the city.

Based on the assessment of PM emission sources and the estimate of PM emissions in Karachi, this chapter also presents the following additional conclusions and guidance:

- Effectively enforce the banning of solid waste burning in the city and find solutions to the issue of recycling so that trucks stop selling/dumping waste at unofficial sites where waste is burned to separate out recyclable materials.
- Move any existing brick kilns, metal foundries, and scrap smelters out of the city and consider wind directions and future urban development when deciding on acceptable locations.
- Improve street cleaning to reduce resuspension of road dust and implement measures to reduce construction dust.
- Control emissions from large point sources (ensure that existing control equipment is operating properly; install new control equipment where none presently exists), but assess on a case-by-case basis in light of population exposure, location, and wind directions.

Effective implementation of a vehicle emission inspection and maintenance (I&M) program is also an important action to reduce vehicle emissions. A starting point may be to first subject heavy-duty diesel vehicles to more stringent maintenance regimens. This would include large public buses and trucks operating in urban areas.

Controlling emissions from gasoline vehicles is also important. Many of the pollutants from combustion of gasoline have health effects. Moreover, some of the emissions from gasoline vehicles form into secondary particulates such as sulfates and nitrates. Thus, gasoline must be of a quality that allows new and imported secondhand vehicles entering the roads to comply with more stringent emission standards.

The government is already pursuing many of the above-mentioned actions at various degrees, such as the conversion of rickshaws, banning of two-stroke vehicles, selective use of low-sulfur fuel oil, and banning of waste burning in the city. The analysis in this chapter provides support for these actions as well as highlighting additional ones.

To improve the assessment of benefits and costs of interventions for better PM ambient air quality in Karachi, and to develop well-targeted interventions, more PM apportionment and emission inventory studies are needed. This will enhance the understanding of the total PM emission load and the contribution of sectors to urban air pollution in Karachi.

Notes

1. The data on emissions sources and inventories in Pakistan need to be improved to refine the design and targeting of control strategies for both mobile and stationary sources.

2. $PM_{2.5}$ and PM_{10} are particulates with a diameter smaller or equal to 2.5 and 10 micrometers (μm), respectively.

3. Alam and others (2011) report a $PM_{2.5}/PM_{10}$ ratio of 0.48 from monitoring at SUPARCO and M. A. Jinnah Road in 2010. SUPARCO reported PM_{10} concentrations of 194 μg/m^3 in Karachi in 2004.

4. The area of Karachi is 2,950 km^2 without the Kirther National Park Area in Gadap.

5. Excluding Gadap (4), Kemari (6), Bin Qasim (10), and the cantonments in figure 2.4.

6. See WHO (2004) for a description of the model and data applied for estimation of annual average PM_{10} concentrations in cities in the world with a population of more than 100,000.

7. Estimated ALRI and cardiopulmonary mortality from PM pollution has been adjusted to reflect the fact that several environmental risk factors cause an increase in the same disease in a particular age group of the population. In addition to urban air pollution, inadequacies in water supply, sanitation, and hygiene also contribute to ALRI mortality through child malnutrition, while road traffic noise contributes to heart disease and cerebrovascular disease mortality.

8. Estimates follow the methodology of WHO using age weighting and a 3% discount rate.

9. Intelligence quotient (IQ) is a score on standardized tests designed to assess intelligence.

10. An exception is a study in a low-income urban area in eastern Karachi that found an average BLL of 11.7 μg/dL among adults 18–60 years of age (Yakub and Iqbal 2010).

11. None of these studies reports the year of BLL measurements.

12. The standard deviation (SD) of BLLs applied in the distribution function is the value that gives a distribution with 20% of urban children having a BLL > 10 μg/dL and 10% of rural children having a BLL > 10 μg/dL. About 63% of children under five in Sindh live in rural areas and 37% live in urban areas (GoP 2009).

13. A study in 25 rural villages in Uttar Pradesh in India in 2009 found an average lead concentration of 74 μg/L in drinking water from hand pumps with a mean depth of over 30 meters (Chaudhary and Sharma 2010).

14. The study is undated but published in 2005 or more recently. The study does not state the year of the groundwater sampling.

15. IQ losses associated with BLLs > 10 μg/dL have been established long ago.

16. Annual loss of IQ points is calculated as Δ IQ/5 by assuming that the children's IQ points are lost in the first five years of life.

17. The high bound reflects the estimated loss in income for males and females in Salkever (1995), weighted by the labor force participation rates of males and females in Sindh province reported by FBS (2010). The low and high bounds do not include the effect of IQ on the rate of participation in the labor force.

18. Tons of secondary particulates are estimated based on Mansha and others (2011).

19. The lower bound reflects valuation of mortality risk using the human capital approach. The upper bound reflects valuation of mortality risk using a value of statistical life (VSL). This book applies the upper bound in the benefit-cost analysis (CBA).

20. A cost of a DOC of US$1,500 was applied for heavy-duty trucks and large buses, and a cost of US$1,000 was applied for minibuses and light-duty vans. A discount rate of 10% was applied to annualize the cost of the DOC.

21. A discount rate of 10% was applied to annualize the cost of conversion to CNG.

22. Cost is adjusted by inflation from 2005 to 2010.

23. A discount rate of 10% is applied to motorcycles and rickshaws to annualize engine and conversion costs.

24. Benefit-cost ratios for PM emission-control interventions in this section are based on midpoint costs and the ranges of vehicle usage that this book estimates.

References

ADB (Asian Development Bank). 2005. *Karachi Mega-Cities Preparation Project*. Final report. Vol. 1. TA 4578-Pakistan.

Ahamed, M., S. Verma, A. Kumar, and M. K. Siddiqui. 2010. "Blood Lead Levels in Children of Lucknow, India." *Environmental Toxicology* 25 (1): 48–54.

Ahmad, T., A. Mumtaz, D. Ahmad, and N. Rashid. 2009. "Lead Exposure in Children Living Around the Automobile and Battery Repair Workshops." *Biomedica* 25: 128–32.

Alam, K., T. Blaschke, P. Madl, A. Mukhtar, M. Hussain, T. Trautmann, and S. Rahman. 2011. "Aerosol Size Distribution and Mass Concentration Measurements in Various Cities of Pakistan." *Journal of Environmental Monitoring* 13: 1944–52.

Alberini, A., and A. Krupnick. 2000. "Cost-of-Illness and Willingness-to-Pay Estimates of the Benefits of Improved Air Quality: Evidence from Taiwan." *Land Economics* 76: 37–53.

Bloomberg. 2010. http://www.bloomberg.com/news/2010-06-11/low-sulfur-fuel-oil-premium-togain-on-environmental-rules-energy-markets.html.

Blumberg, K., K. He, Y. Zhou, H. Liu, and N. Yamaguchi. 2006. "Costs and Benefits of Reduced Sulfur in China." The International Council on Clean Transportation. December 2006.

Blumberg, K., M. Walsh, and C. Pera. 2003. *Low-Sulfur Gasoline and Diesel: The Key to Lower Vehicle Emissions*. http://www.unep.org/pcfv/PDF/PubLowSulfurPaper.pdf.

Canfield, R. L., C. R. Henderson, D. A. Cory-Slechta, C. Cox, T. A. Jusko, and B. P. Lanphear. 2003. "Intellectual Impairment in Children with Blood Lead Concentrations below 10 μg per Deciliter." *New England Journal of Medicine* 348 (16): 1517–26.

Carlisle, J., K. Dowling, D. Siegel, and G. Alexeeff. 2009. "A Blood Lead Benchmark for Assessing Risks from Childhood Lead Exposure." *Journal of Environmental Science and Health Part A* 44 (12): 1200–08.

CDGK (City District Government Karachi). 2007. *Karachi Strategic Development Plan 2020*. Master Plan Group of Offices, Karachi, Pakistan.

Chaudhary, V., and M. K. Sharma. 2010. "Risk Assessment of Children's Blood Lead Level in Some Rural Inhabitations in Western Uttar Pradesh, India." *Toxicological and Environmental Chemistry* 92 (10): 1929–37.

Cropper, M., and W. Oates. 1992. "Environmental Economics: A Survey." *Journal of Economic Literature* 30: 675–740.

Daily Times. 2012. http://www.dailytimes.com.pk/default.asp?page=2012%5C01%5C16%5Cstory_16-1-2012_pg7_26.

Dickie, M., and S. Gerking. 2002. "Willingness to Pay for Reduced Morbidity." Paper presented at Economic Valuation of Health for Environmental Policy: Assessing Alternative Approaches, Orlando, FL, March 18–19.

ECON. 2006. "Urban Air Pollution Control in Peru." Prepared for the Peru Environmental Analysis, World Bank, ECON Analysis, Oslo. http://www.google.com/url?sa=t&source=web&cd=1&ved=0CCAQFjAA&url=http%3A%2F%2Fecon.no%2Fstream_file.asp%3FiEntityId%3D3695&rct=j&q=Urban%20Air%20Pollution%20Control%20in%20Peru&ei=lhQ-TuimMKvUiAKi6JXDBg&usg=AFQjCNGZ_11jUKHrzqB1k_v-lW38GYdYLA&cad=rja.

EIA (Energy Information Administration). 2012. *Petroleum Marketing Monthly* (February). EIA, Department of Energy, Washington, DC.

Enstrat International Ltd. 2003. *Cost of Diesel Fuel Desulphurisation for Different Refinery Structures Typical of the Asian Refinery Industry.* Final report prepared for the Asian Development Bank, January 10, 2003.

FBS (Federal Bureau of Statistics). 2009. *Household Integrated Economic Survey 2007–08.* FBS, Pakistan.

———. 2010. *Labor Force Survey 2008–09.* FBS, Pakistan.

Fewtrell, L., R. Kaufmann, and A. Pruss-Ustun. 2003. *Lead: Assessing the Environmental Burden of Disease at National and Local Levels.* Environmental Burden of Disease Series, No. 2. Geneva: WHO.

Ghauri, B. 2008. "Satellite Data Applications in Atmospheric Monitoring." PowerPoint presentation presented at the United Nations/Austria/European Space Agency Symposium, Graz, Austria, September 9–12.

———. 2010. *Institutional Analysis of Air Quality Management in Urban Pakistan.* Study commissioned by the World Bank. Washington, DC: World Bank. http://cleanairinitiative.org/portal/system/files/.../AQM_Draft_Final_Report.pdf.

Gilbert, S. G., and B. Weiss. 2006. "A Rationale for Lowering the Blood Lead Action Level from 10 to 2 µg/dL." *NeuroToxicology* 27: 693–701.

GoP (Government of Pakistan). 2009. *Pakistan Demographic Survey 2007.* Statistics Division, Government of Pakistan.

Gould, E. 2009. "Childhood Lead Poisoning: Conservative Estimates of the Social and Economic Benefits of Lead Hazard Control." *Environmental Health Perspectives* 117 (7): 1162–67.

Grosse, S. D., T. D. Matte, J. Schwartz, and R. J. Jackson. 2002. "Economic Gains Resulting from the Reduction in Children's Exposure to Lead in the United States." *Environmental Health Perspectives* 110 (6): 563–69.

Gwilliam, K., M. Kojima, and T. Johnson. 2004. "Reducing Air Pollution from Urban Passenger Transport." World Bank, Washington, DC. http://www-wds.worldbank.org/external/default/WDSContentServer/IW3P/IB/1998/11/17/000178830_98111703524419/Rendered/PDF/multi_page.pdf.

Junejo, S. A. n.d. "Groundwater Quality in Sindh." Indus Institute for Research and Education (IIRE), Hyderabad, Pakistan.

Jusko, T. A., C. R. Henderson, B. P. Lanphear, D. A. Cory-Slechta, P. J. Parsons, and R. L. Canfield. 2008. "Blood Lead Concentrations < 10 µg/dL and Child Intelligence at 6 Years of Age." *Environmental Health Perspectives* 116 (2): 243–48.

Kadir, M. M., N. Z. Janjua, S. Kristensen, Z. Fatmi, and N. Sathiakumar. 2008. "Status of Children's Blood Lead Levels in Pakistan: Implications for Research and Policy." *Public Health* 122 (7): 708–15.

Khan, D. A., W. M. Ansari, and F. A. Khan. 2011. "Synergistic Effects of Iron Deficiency and Lead Exposure on Blood Lead Levels in Children." *World Journal of Pediatrics* 7 (2): 150–54.

Khan, D. A., S. Qayyum, S. Saleem, W. M. Ansari, and F. A. Khan. 2010. "Lead Exposure and Its Adverse Effects among Occupational Worker's Children." *Toxical and Industrial Health* 26 (8): 497–504.

Lanphear, B. P., R. Hornung, J. Khoury, K. Yolton, P. Baghurst, D. C. Bellinger, R. L. Canfield, K. N. Dietrich, R. Bornschein, T. Greene, S. J. Rothenberg, H. L. Needleman, L. Schnaas, G. Wasserman, J. Graziano, and R. Roberts. 2005. "Low-level Environmental Lead Exposure and Children's Intellectual Functions: An International Pooled Analysis." *Environmental Health Perspectives* 113 (7): 894–99.

Larsen, B. 2005. "Cost-benefit Analysis of Environmental Protection in Colombia." Background paper prepared for the Colombia Country Environmental Analysis. Colombia: Mitigating Environmental Degradation to Foster Growth and Reduce Inequality, Washington, DC: World Bank.

———. 2007. "Cost-Benefit Analysis of Environmental Health Interventions in Urban Greater Dakar, Senegal." Paper prepared for the Senegal Country Environmental Analysis, World Bank, Washington, DC. Prepared from ECON/Roche Canada.

———. 2009. "Cost-Benefit Analysis of Selected Environmental Health Interventions: International Evidence and Applications to the Philippines." Paper prepared for the Philippines CEA, World Bank, Washington, DC.

———. 2011. "Benefits and Costs of Select Environmental Health Interventions in the MENA Countries." Paper prepared for the World Bank/CMI, Washington, DC.

Larsen, B., G. Hutton, and N. Khanna. 2009. "Air Pollution." In *Global Crisis, Global Solutions: Costs and Benefits*, edited by B. Lomborg. Cambridge, U.K.: Cambridge University Press.

Liu, J., Y. Ai, L. McCauley, J. Pinto-Martin, C. Yan, X. Shen, and H. Needleman. 2011. "Blood Lead Levels and Associated Sociodemographic Factors among Preschool Children in the South Eastern Region of China." *Pediatric and Perinatal Epidemiology* 26 (1): 61–69.

Mansha, M., B. Ghauri, S. Rahman, and A. Amman. 2011. "Characterization and Source Apportionment of Ambient Air Particulate Matter ($PM_{2.5}$) in Karachi." *Science of the Total Environment* 425:176–83.

MECA (Manufacturers of Emission Controls Association). 2009. *Retrofitting Emissions Controls on Diesel-Powered Vehicles*. Washington, DC: MECA. http://www.meca.org.

Meszler, M. 2007. "Air Emissions Issues Related to Two- and Three-Wheeled Motor Vehicles: An Initial Assessment of Current Conditions and Options for Control." International Council of Clean Transportation. http://www.theicct.org/documents /0000/1021/Meszler_2_3Wheelers_2007_v3.pdf.

Muennig, P. 2009. "The Social Costs of Childhood Lead Exposure in the Post-Lead Regulation Era." *Archives of Pediatric and Adolescent Medicine* 163 (9): 844–49.

Nichani, V., W. I. Li, M. A. Smith, G. Noonan, M. Kulkarni, M. Kodavor, and L. P. Naeher. 2006. "Blood Lead Levels in Children after Phase-Out of Leaded Gasoline in Bombay, India." *The Science of the Total Environment* 363 (1–3): 95–106.

Pichery, C., M. Bellanger, D. Zmirou-Navier, P. Glorennec, P. Hartemann, and P. Grandjean. 2011. "Childhood Lead Exposure in France: Benefit Estimation and Partial Cost-Benefit Analysis of Lead Hazard Control." *Environmental Health* 10: 44.

Rahbar, M. H., F. White, M. Agboatwalla, S. Hozhabri, and S. Luby. 2002. "Factors Associated with Elevated Blood Lead Concentrations in Children in Karachi, Pakistan." *WHO Bulletin* 80 (10): 769–75.

Reuters. 2012. http://www.reuters.com/article/2012/02/27/markets-europe-distillatesid USL5E8DRADZ20120227.

Riddell, T. J., O. Solon, S. A. Quimbo, C. M. C. Tan, E. Butrick, and J. W. Peabody. 2007. "Elevated Blood-Lead Levels among Children Living in the Rural Philippines." *WHO Bulletin* 85 (9): 674–80.

Rothenberg, S., and J. Rothenberg. 2005. "Testing the Dose-Response Specification in Epidemiology: Public Health and Policy Consequences for Lead." *Environmental Health Perspectives* 113 (9): 1190–95.

Roy, A., D. Bellinger, H. Hu, J. Schwartz, A. S. Ettinger, R. O. Wright, M. Bouchard, K. Palaniappan, and K. Balakrishnan. 2009. "Lead Exposure and Behavior among Young Children in Chennai, India." *Environmental Health Perspectives* 117 (10): 1607–11.

Salkever, D. S. 1995. "Updated Estimates of Earnings Benefits from Reduced Exposure of Children to Environmental Lead." *Environmental Research* 70: 1–6.

Sánchez-Triana, E., S. Enriquez, B. Larsen, and E. Golub. 2014. *Environmental and Climate Change Priorities for the Sindh Province*. Environment and Development Series. Washington, DC: World Bank.

Sara Group. 2008. "IEUBK Lead Modeling Overview and Results." Appendix Q in Sudbury Area Risk Assessment. Volume 2. http://www.sudburysoilsstudy.com/EN /media/Volume_II/Volume_II_Report/SSS_Vol_II_HHRA_Appendix_Q_IEUBK _LeadModellingOverviewandResults_FinalReport_021408.pdf. Accessed July 31, 2011.

Schwartz, J. 1994. "Societal Benefits of Reducing Lead Exposure." *Environmental Research* 66: 105–12.

Sindh EPA. 2010. Daily $PM_{2.5}$ data collected from air quality monitoring network Karachi. Unpublished.

SNC-Lavalin. 2011. "National Power System Expansion Plan 2011–30." Prepared by SNC-Lavalin International Inc. in association with National Engineering Services Pakistan (Pvt.) Ltd. for the National Transmission and Despatch Company Ltd.

Stevens, G., A. Wilson, and J. Hammitt. 2005. "A Benefit-Cost Analysis of Retrofitting Diesel Vehicles with Particulate Filters in the Mexico City Metropolitan Area." *Risk Analysis* 25 (4): 883–99.

Surkan, P. J., A. Zhang, F. Trachtenberg, D. B. Daniel, S. McKinle, and D. C. Bellinger. 2007. "Neuropsychological Function in Children with Blood Lead Levels < 10 µg/dL." *NeuroToxicology* 28 (6): 1170–77.

Toxic Link. 2010. "Toxic Trinkets: An Investigation of Lead in Children's Jewelry in India." New Delhi, India: Toxic Link.

Ul-Haq, N., M. A. Arain, N. Badar, M. Rasheed, and Z. Haque. 2011. "Drinking Water: A Major Source of Lead Exposure in Karachi, Pakistan." *Eastern Mediterranean Health Journal* 17 (11): 882–86.

UNEP (United Nations Environment Programme). 2006. *Opening the Door to Cleaner Vehicles in Developing and Transition Countries: The Role of Lower Sulphur Fuels*. UNEP Partnership for Clean Fuels and Vehicles, Nairobi.

————. 2009. *Cleaning Up Urban Bus Fleets with Focus on Developing and Transition Countries*. Nairobi: UNEP.

————. 2011. *Asia-Pacific 50 ppm Diesel Sulphur Matrix*. Nairobi: UNEP.

USEPA (United States Environmental Protection Agency). 2012. http://www.epa.gov /cleandiesel/technologies/retrofits.htm.

WHO (World Health Organization). 2004. *Comparative Quantification of Health Risks: Global and Regional Burden of Disease Attributable to Selected Major Risk Factors*. Geneva, Switzerland: WHO.

Wilson, C. 2003. "Empirical Evidence Showing the Relationships between Three Approaches for Pollution Control." *Environmental and Resource Economics* 24: 97–101.

World Bank. 2006. "Pakistan Strategic Country Environmental Assessment." World Bank, Washington, DC.

————. 2011. *World Development Indicators*. Washington, DC: World Bank.

World Gazetteer. 2011. Population data and statistics. www.world-gazetteer.com.

Yakub, M., and M. P. Iqbal. 2010. "Association of Blood Lead (Pb) and Plasma Homocysteine: A Cross-Sectional Survey in Karachi, Pakistan." *PLoS One* 5 (7): e11706.

CHAPTER 5

Industrial and Other Stationary Sources

Introduction

Economic growth and industrialization have gone hand in hand in Pakistan, but these processes have not been linear and they have been accompanied by significant negative externalities, particularly air pollution. Services represented about 53% of Pakistan's gross domestic product (GDP) in 2008 and grew at an average annual rate of around 6% during 1998–2008. In comparison, industry's share was 27% of GDP and grew at an average rate of around 7% during the same period, while agriculture amounted to 20% of GDP and grew at a yearly average rate of 3%. Current projections indicate that services will remain the economy's most important sector in the future and that industry's contribution will increase as agriculture's share falls (Pakistan Planning Commission 2011).

Pakistani industry is outdated and risks losing markets at a time when it may have the opportunity to occupy the space being left by industrial giants like China. Industries tend to concentrate in a few geographic locations where their competitiveness is enhanced by the availability of specialized labor, inter-industry spillovers, higher road density, local transfer of knowledge, and access to international supplier and buyer networks. These factors largely explain the clustering of large-scale manufacturing and the high employment levels around the metropolitan areas of Karachi and Lahore.

While agglomeration economies exert a strong pull factor on industrialization, if not well managed, air pollution and other "public bads" associated with this process (for example, time and congestion, and crime and urban violence) may dissipate the benefits of economic growth. The concern is that agglomeration economies in Pakistan may reach their limits early in the industrialization process.

There are a number of signs that this problem is already materializing (Burki and others 2011). As recognized in the Government of Pakistan's (GoP) *Framework for Economic Growth* (2011), industrial growth offers opportunities to support sustained economic growth in Pakistan (Pakistan Planning Commission 2011). However, the potential associated with these opportunities will not be realized without specific governmental interventions.

Air quality management (AQM) will play a key role in realizing the potential noted above. Enforcement of environmental standards and adoption of cleaner production practices are among the measures that can generate substantial benefits for firms while also contributing to improve environmental conditions in Pakistan (for example, cost savings from more efficient use of resources such as energy, and access to export markets in which poor industrial environmental performance can be a barrier to entry).

One of the striking features of Pakistan's economic structure is the way in which it departs from that of developing countries with strong emerging economies. In comparison to China, Indonesia, and Malaysia, Pakistan has the largest agricultural share of GDP and the smallest industrial share of GDP. In addition, in comparison to India, Sri Lanka, and Mexico, Pakistan has the largest agricultural share of GDP. These results are surprising given Pakistan's urbanization trend (figure 5.1).

Industrial activities, particularly those using fossil fuels, are a significant source of air pollution. Thermal power plants, sugar mills, and industrial establishments from various sectors (textiles, tanneries, paper, pharmaceutical, cement, and fertilizer) use furnace oil that is high in sulfur content and contributes significant amounts of air pollutants. A wide range of small- to medium-scale industries (including brick kilns, steel re-rolling, steel recycling,

Figure 5.1 Economic Structure of Pakistan and other Developing Countries

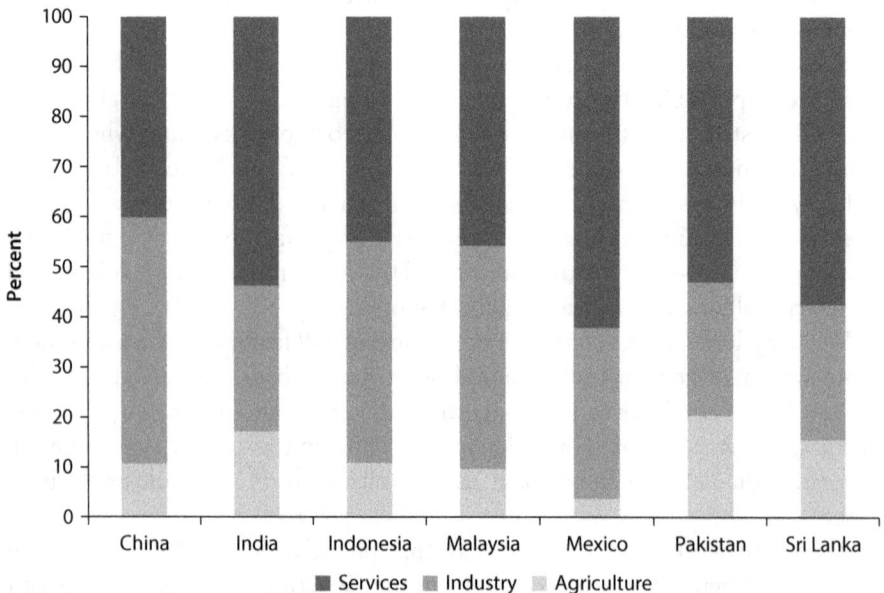

Source: CIA 2008.

and plastic molding) use dirty "waste" fuels, such as old tires, paper, wood, and textile waste, that generate air pollutants. Nonpoint sources, including the burning of agricultural residues, of sugarcane fields, and of municipal waste also contribute to air pollution. The arid conditions in some areas in Pakistan, compounded by dust storms in the summer months, result in clouds of fine dust that form a haze over many cities. The Pakistan Environmental Protection Agency (Pak-EPA 2010) and Ghauri (2010) also have identified transboundary pollution from stationary sources as a significant source of particulate matter (PM).

This chapter analyzes stationary sources and identifies interventions to control air pollution from industrial and other stationary sources. The section on "Inventory of Stationary Sources" identifies the main stationary sources of air emissions. The "Industrial Air Pollution" section analyzes the environmental profile of the main industrial sectors in Pakistan with a closer look at their specific air pollution issues. The section on "Control of Air Pollution from Stationary Sources" presents an assessment of the main cleaner production initiatives implemented in each sector, and the "Control of Diffuse Sources" section identifies main gaps and the way forward in the adoption of further measures for cleaner production with regard to air quality.[1]

Inventory of Stationary Sources

Stationary sources of air pollutant emissions in Pakistan include thermoelectric power plants, textiles, tanneries, sugar mills, cement, fertilizer, paper, brick kilns, and burning of agricultural residues, of sugarcane fields, and of municipal waste. According to the Census of Manufacturing Industries, the number of industrial facilities in Pakistan amounts to 6,417 industrial units. Textiles, food products and beverages, tanneries, and chemical industries account for 60% of the total (table 5.1 and figure 5.2). Most of the industrial facilities are located in industrial clusters in Punjab and Sindh.

Most data on the emission of pollutants by stationary sources, such as factories, come from a database for self-reported pollution information. The federal government issued the National Environmental Quality Standards (NEQS) or the Self-Monitoring and Reporting Tool (SMART) in 2001, which instructs all the industrial units to submit reports as per the requirements of industry categories in which their unit is placed. In the case of gaseous emissions, industrial units are divided into two categories. Category A industries are instructed to submit SMART reports monthly, and the reporting requirement for category B industries is quarterly. Priority parameters have been selected for different industrial sectors with respect to the type of pollution they generally discharge and emit. Separate forms have been developed for reporting the liquid and gaseous emissions monitoring. The test reports from a certified environmental laboratory are required to be annexed to the monitoring report. Sample testing and analysis should be conducted as per the Environmental Sample Rules, 2001. User-friendly software was developed for

Table 5.1 Industrial Units of Pakistan by Sector, 2005

Sector	Industrial establishments
Food products and beverages	1,861
Tobacco	13
Manufacturing of textiles	1,328
Wearing apparel	326
Leather products	142
Wood and wood products	62
Paper and paper products	133
Publishing, printing, and reproduction	47
Coke and petroleum	31
Chemicals and chemical products	493
Rubber and plastic products	170
Other non-metallic mineral products	482
Basic metals	291
Fabricated metal products	144
Machinery and equipment	372
Electrical machinery and apparatus	67
Radio, TV, and communication equipment	14
Medical and optical instruments	95
Motor vehicles and trailers	139
Other transport equipment	47
Furniture	130
Recycling	30
Total	6,417

Source: FBS 2005–06, 17.

reporting the data in electronic form. The data so entered could be sent to the respective Environmental Protection Agency (EPA) via email.

The voluntary SMART program implemented to identify stationary sources and monitor industrial emissions has proven unable to provide information on emission of pollutants by stationary sources in Pakistan. The level of reporting on air pollutant emissions is low. Only 99 out of 6,417 industrial facilities have registered their emissions under the SMART Program (table 5.2). In line with experiences in other countries, the Pakistan voluntary agreement to disclose emissions of air pollutants has proven ineffective (Esty and Porter 2001; Khan 2010; Laiq 2011; Morgenstern and Pizer 2007).

Voluntary agreements in most countries have been ineffective. This is corroborated by research findings in the United States and the European Union, where voluntary agreements "… may represent a shift in emphasis from the worst polluters to those most willing to abate on their own initiative. Some see voluntary programs as a distraction from the real work of taking mandatory action" (Morgenstern and Pizer 2007, 3). At some point, country conditions may have dictated a transition from no reporting to voluntary reporting; however, there is evidence this is a suboptimal situation that contributes little to disclosing emissions or abating pollution.

Figure 5.2 Industrial Units of Pakistan by Sector and Region, 2005

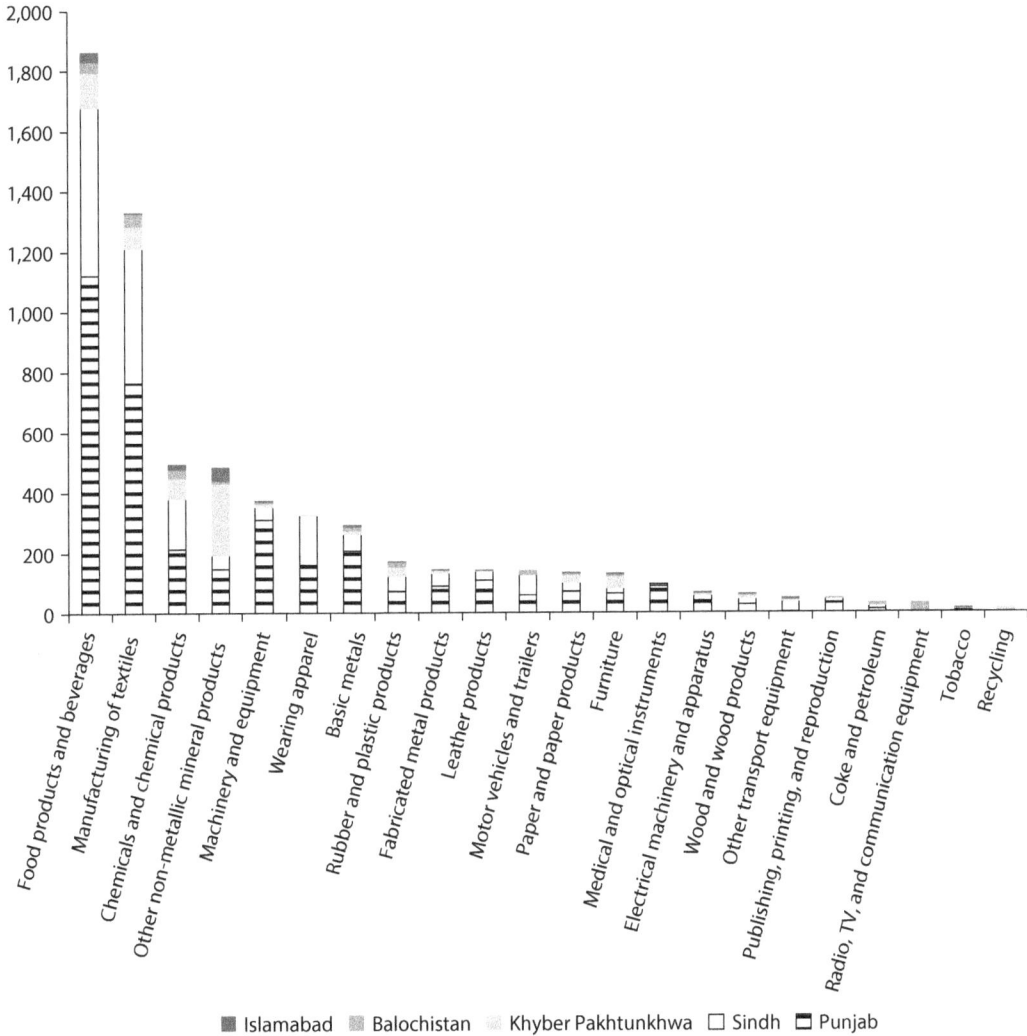

Legend: ■ Islamabad ▨ Balochistan ▨ Khyber Pakhtunkhwa □ Sindh ▤ Punjab

Source: FBS 2005–06, 17.

Industrial Air Pollution

Industrial activities, particularly those using fossil fuels, are a significant source of air pollution. Thermal power plants, textile, tanneries, paper, pharmaceutical, cement, fertilizer, and sugar mills use furnace oil that is high in sulfur content and contributes significant amounts of air pollutants. Energy generation in those industrial activities adds air pollutants because of the poor maintenance of boilers.

Industrial emissions are further compounded by the widespread use of small diesel and petrol-electric generators in commercial and residential areas in response to the poor reliability of electricity supplies. A wide range of small- to

Table 5.2 SMART Registration and Reporting by Industry Sector and Provinces, 2010

Plant type	Provincial category type of plants					
	Federal	Punjab	Sindh	KP	Balochistan	Total
Cement	0	3	0	2	0	5
Chlor-alkali (mercury and diaphragm cell)	0	1	0	0	0	1
Dairy	0	2	0	0	0	2
Industrial chemicals	0	2	5	0	0	7
Iron and steel	0	1	1	0	0	2
Leather tanning	0	1	0	0	0	1
Nitrogenous fertilizers	0	3	2	0	0	5
Oil and gas production/thermal power plant	0	0	17	2	2	21
Other plant types	1	2	7	2	0	12
Pesticide formulation	0	1	2	0	0	3
Petroleum refining	0	1	2	0	0	3
Pharmaceutical	0	0	1	0	0	1
Pharmaceutical formulation	1	4	9	1	0	15
Plastic material and products	0	0	2	0	0	2
Printing	0	1	1	0	0	2
Sugar	0	2	0	0	0	2
Synthetic fiber	0	1	0	1	0	2
Textile	0	2	1	0	0	3
Textile processing	0	3	2	0	0	5
Thermal power plants (gas-fired)	0	1	0	0	0	1
Thermal power plants (oil- and coal-fired)	0	4	0	0	0	4
Total	2	35	52	8	2	99

Source: http://www.environment.gov.pk/smart/Site/Memebers.html.
Note: SMART = Self-Monitoring and Reporting Tool; KP = Khyber Pakhtunkhwa.

medium-scale industries (including brick kilns, steel re-rolling, steel recycling, and plastic molding) cause a disproportionate share of pollution through their use of dirty "waste" fuels, such as old tires, paper, wood, and textile waste. Although data are limited, sporadic monitoring of air pollutants in industrial areas in Pakistan suggests that international standards for particulate matter of less than 10 microns (PM_{10}) and nitrogen oxides (NO_x) are exceeded frequently (table 5.3).

Key binding constraints to controlling air pollution in Pakistan include gaps in identification of stationary sources, an ineffective voluntary program for monitoring air emissions from stationary sources, a weak regulatory framework, and lack of enforcement of regulations.

In the industrial sector, inefficient boilers and generators generate PM and sulfur oxides (SO_x) and NO_x. High-sulfur content of furnace oil contributes significantly to PM and sulfur dioxide (SO_2) emissions.

Sulfur concentrations in furnace oil used in Pakistan electricity generation and industrial applications range from 2,000 to 9,000 parts per million (ppm). There is a need to reduce the sulfur content of furnace oil. In addition, subsidies result in lower consumer prices and discourage efficient gas use. Gas for feedstock to fertilizer plants was sold at one-third of the average end-user tariff.

Table 5.3 Air Quality in Industrial Areas of Pakistan
$\mu g/m^3$

Area	Year	Average	Source	TSP	PM_{10}	NO_2	NO	NO_x
Korangi industrial area	1999	24 h	Hashmi and Khani 2003	–	147	–	–	–
Lahore industrial area (undefined)	2002–03	24 h	FBS 2004	246	–	–	29.4 (ppb)	44 (ppb)
Cunian/Kasur. Multan Road	2002–03	24 h		265	–	–	–	65 (ppb)
Faisalabad. Samundri Road		24 h		471	–	29.1 (ppb)	–	41.2 (ppb)
Gujranwala. Sheikupura Road		24 h	FBS 2004	650	–	35.9 (ppb)	–	65.1 (ppb)
Lahore. Qartaba Chowk		24 h		623	–	–	58 (ppb)	214 (ppb)
R. Y. Khan. industrial area		24 h		345	–	–	42 (ppb)	79 (ppb)

Sources: FBS 2004; Hashmi and Khani 2003; Rajput and others 2005; Smith and others1996; Wasim and others 2003; WHO/UNEP 1992.
Note: – = data not collected; PM_{10} = particulate matter of less than 10 microns; TSP = total suspended particle; NO_2 = nitrogen dioxide; NO = nitric oxide; NO_x = nitrogen oxide.

Table 5.4 Gas Prices in Pakistan

Consumer category	Sales prices as of July 1, 2010 (rounded figures)		
	PRs./mmBtu	US$/mmBtu	Selected cross subsidies[a] US$ Million
Residential	152 (average)	1.77	397
Commercial	464	5.42	—
Cement	536	6.26	—
Fertilizer (feedstock)	102	1.19	389
Industry	382	4.46	—
Transport (CNG)	504	5.88	—
Power generation	394	4.60	—
Independent power producers	332	3.88	—
Captive power	382	4.46	—
Weighted average gas price	321.5	3.75	—

Source: Oil and Gas Regulatory Agency 2011.
Note: Large industrial consumers have individual prices. — = not available; CNG = compressed natural gas.
a. Cross subsidy calculation based on these gas volumes: residential: 214 bcf; fertilizer feedstock: 162 bcf; total market: 1,269 bcf for fiscal year 2009 (Ministry of Petroleum and Natural Resources 2009). The calculation does not take into account variations in cost of service. A calorific value of 930 btu/cf was used.

Average end-user tariff for gas in FY 2010 was US$3.75/mmBtu, and the average tariff to households was half that level due to cross-subsidies from rates charged to industrial consumers and power plants (table 5.4). The industrial sector consumes the largest share of energy in Pakistan and its consumption has grown significantly over recent years (figure 5.3).

Textiles constitute the largest industrial sector of Pakistan with respect to production, export, and labor force. The textile sector has an installed capacity of 1,922 million kilograms of yarn spinning, and contributes to around 8% of the country's GDP, 55% of total exports, and 46.6% of the labor force in the manufacturing sector (Ministry of Finance 2010; Ministry of Textile Industry 2009; TDAP 2008). Textile-processing units are mostly located in Lahore, Karachi, and Faisalabad. In the Lahore and Faisalabad regions, most of the units are small and

Figure 5.3 Energy Consumption by Sector, 2005–12
Tonnes of oil equivalent

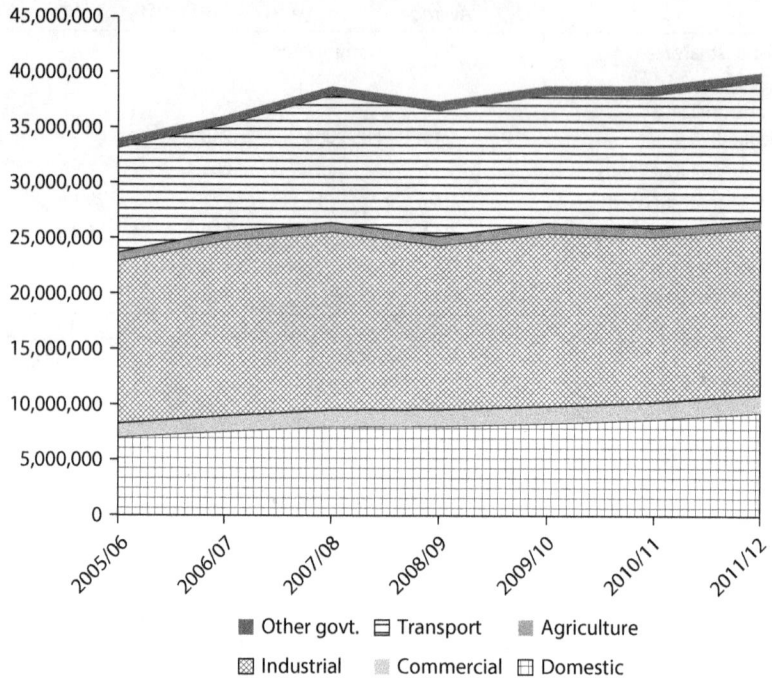

Source: Ministry of Petroleum and Natural Resources 2013.

Table 5.5 Typical Emission for Textile Processing Units

Emission (mg/nm³)	Sources of emissions					
	Boiler	Generator	Thermoil heater	Stenter	Color kitchen	Screen development
Nitrogen oxides (NO$_x$)	40–90	800–1,700	40–90	–	–	–
Carbon monoxide (CO)	20–800	600–1,400	200–500	–	–	–
Sulfur oxides (SO$_x$)	–	400–1,200	80–100	–	–	–
Volatile organic compounds (VOC)	–	–	–	50–150	–	70–160
Ammonia (NH$_3$)	–	–	–	–	20–65	–

Source: Cleaner Technology Program for Textile Industry (CTPT) database by Khan 2010.

medium sized, whereas large units are mostly located in Karachi and Faisalabad (Khan 2010). Textile plants emit a number of air pollutants (table 5.5).

By 2008, the leather sector contributed about 6% of Pakistan's total exports (TDAP 2008). Major export items in this sector are finished leather, leather garments, and gloves. The main buyers of Pakistani products are China, France, Germany, Italy, Spain, Turkey, the United Arab Emirates, the United Kingdom, and the United States. Major clusters of tanneries are located in Karachi, Kasur, Lahore, Sheikhupura, Gujranwala, Multan, and Sialkot.

For leather production, locally available raw material (hides and skins) and imported process chemicals are used (Khan 2010). Concentrations of PM are high due to dust emissions from shaving and buffing machines. According to Khan (2010), some medium and large tanneries have installed dust collectors.

Pakistan is one of the largest producers of refined sugar. The sugar industry in Pakistan generates 1.9% of GDP. By 2009, 87 large sugar mills were located in Pakistan, mainly in rural areas of Punjab and Sindh (Khan 2010). The total crushing capacity of sugar industry in Pakistan is approximately 500,000 tonnes of cane per day (Khan 2010). PM containing fly ash and unburned bagasse particles found in the boilers' flue gases are the major issues in most of the sugar mills.[2] According to Khan (2010), only one sugar mill claimed to comply with air emissions regulations.

During 2008–09, the production of paper and board sector was about 530,000 tonnes (Ministry of Finance 2010). In pulp and paper mills, wheat straw cleaning cyclones, the chemical house, and utilities constituted the major sources of PM, NO_x, and SO_x. Main sources of the air pollutants in paper processing factories included boilers and continuous digester blow tanks (Khan 2010).

Pakistan's fertilizer industry has a production capacity of approximately 4,259 product tonnes per annum (Ministry of Finance 2010). Natural gas is used as feedstock, as well as fuel in the manufacturing of nitrogen fertilizer. The consumption of gas during 2007–08 was 200,061 mmcft, out of which 80% was used as feedstock and 20 percent as fuel (Khan 2010).

Control of Air Pollution from Stationary Sources

Interventions to control industrial air pollution in Pakistan include substituting gas for fuel oil and coal, introduction of low-sulfur diesel and furnace oil, energy efficiency, and end-of-pipe pollution control technology. Polluting factories could also move out from areas where they violate land-use regulations. Various environmental policy instruments are available to Pakistan for promoting interventions to control industrial air pollution. These include (a) economic-based and market-based instruments, (b) direct regulation by government or "command-and-control" measures, (c) public disclosure, and (d) legal actions. According to economic theory, economic instruments are more efficient for tackling priority environmental problems. In actuality, command and control instruments have been found easier to design and implement and, consequently, more effective for environmental management.

Economic Instruments

Market-based instruments, such as taxes or pricing policies, can be effective in moving existing markets and policies towards improved environmental outcomes. Economic instruments are based on the polluter pays principle (PPP), where the polluting party pays for the damage done to the natural environment. Economic instruments aim at modifying the behavior of economic agents by providing incentives for them to internalize the externalities that they may be producing. Economic instruments include taxes, pollution charges, and tradable permits. A strong system of enforcement is key to enhance the effectiveness of economic instruments (Sánchez-Triana 1993; Sánchez-Triana, Ahmed, and Awe 2007; World Bank 2007).

Pakistan's environmental regulations lack clarity regarding the extent to which industry should bear the costs of pollution reduction. The Pakistan Environment Protection Act (PEPA) of 1997, which established NEQS, does not clearly indicate who is responsible for pollution reduction costs. One result of the vagueness in environmental laws is that the provincial environmental authorities have yet to give out any emission permits. There is also no monitoring system for emissions of stationary sources in Pakistan. The Pakistan Clean Air Program (PCAP) has yet to develop guidelines on how to carry out environmental enforcement. The lack of clarity means that the existing environmental framework of laws, standards, and regulations has, despite successes in other areas, done little to improve the environmental performance of Pakistan's industries (Luken 2009). This failure to stipulate whether industries are responsible for bearing the cost of pollution reduction hinders the implementation of interventions to control industrial air pollution (Luken 2009).

In Pakistan, pollution charges could be designed to promote a shift from using highly polluting fuels such as wood, fuel oil, and diesel to cleaner fuels such as gas. Differentiating prices across fuel types to minimize negative impacts on economic activity requires careful planning. An essential input for fuel taxes or pollution charges design is the targeting of fuels according to pollutants. Fuel types vary in their potential to produce atmospheric emissions that pollute the environment (table 5.6).

Removal of fuel-price distortions could help reduce demand for fuels and improve air quality. Removing price distortions is intended to bring domestic prices in line with international prices and eliminate the difference in tax rates for fuel oil and natural gas. Converting from the use of fuel oil and coal to

Table 5.6 Fuel and Air Pollution Emissions

Fuel type	$PM_{2.5}$	SO_x	NO_x	VOCs
Propane	–	M	L	M
Natural gas	L	M	H	M
Kerosene	M	H	M	M
Diesel (low sulfur)	M	M	H	M
Diesel (high sulfur)	H	H	H	M
Fuel oil	H	H	H	M
Crude oil	H	H	H	M
Coal (low sulfur)	H	M	M	L
Coal (high sulfur)	H	H	M	L
Wood and biomass	H	H	H	H

Source: Adapted from Sánchez-Triana 1993.
Note: H = high concentration of atmospheric emissions; L = low concentration of atmospheric emissions; M = medium concentration of atmospheric emissions; $PM_{2.5}$ = particulate matter of less than 2.5 microns; VOC = volatile organic compound; – = not available.

natural gas can reduce air pollution and provide economic benefits to Pakistan's industrial sectors. Phasing out subsidies to fuels in Pakistan will provide incentives for engaging the industrial sector in international trade of natural gas including imports from the Islamic Republic of Iran or Turkmenistan. High economic growth in 2002–06 caused a surge in energy demand, particularly in the industrial sector. Subsidies to fuel-price distortions and subsidies to electricity tariffs aggravated energy deficits in Pakistan from 2008 onwards (Alahdad 2011). A governmental "tariff differential subsidy" has been established to compensate the energy companies for the inadequate consumer tariffs, but it is neither paid on time nor in full, thereby contributing to "circular debt" where end users do not fully pay their energy service providers, thus choking the financial flow to transmission companies, energy producers, and fuel suppliers. Sector performance is further damaged by large cross subsidies in the natural gas subsector to households and the fertilizer industry distorting consumption, discouraging energy efficiency and conservation measures. By international standards, and compared to oil products, natural gas is inexpensive in Pakistan, which exacerbates the problem of its inefficient use (Kojima 2009).

Command and Control Instruments

Command and control measures can be effectively promoted and aligned with country context and conditions. Command and control regulations focus on preventing environmental problems by mandating standards and technologies to control pollution. This approach generally relies on emissions standards, ambient standards, and technological standards in conjunction with enforcement programs. Typically, parameters for which limits have been established and regulated worldwide include pollutants such as fine particulate matter ($PM_{2.5}$) or coarse particulate matter (PM_{10}) (Gurjar, Molina, and Ojha 2010).

For air pollution control, the most common command and control measures include ambient standards, emission standards, and technology-based and performance-based standards. For example, several countries use command and control instruments for setting ambient air primary standards for $PM_{2.5}$ at 14.0 µg/m^3 (annual average) and 35 µg/m^3 (24-hour average). Examples of common command and control instruments that use technological standards include phased reduction of the sulfur content in furnace oil to 500 ppm in the short term and to 50 ppm or less in the long term. Other technological standards include requiring relocation of industries from areas where ambient air pollution concentrations exceed the legal limits to places with no thermal inversions, good atmospheric mixing conditions, and outside high-population density areas. The relocation of industries has also been used in many countries, such as India and China, as an opportunity to upgrade the plants' technology and improve environmental management in new plants (Li and Bin 2010).

Stringency of air quality standards and strong enforcement are pillars of effective pollution control programs. According to Esty and Porter (2001), out of six generally accepted environmental variables[3] that affect environmental performance and pollution levels in a sample of 42 countries, strict environmental standards have significant power in reducing pollution. Strict environmental standards and enforcement are crucial for decreasing pollution in a country's environmental regulatory regime and strongly correlate with enhanced environmental performance.

Pakistan could start by setting and monitoring standards for PM. Given its impact on human health, Pakistan could start setting $PM_{2.5}$ standards. $PM_{2.5}$, sulfur dioxide (SO_2), nitrate (NO_3), and ozone (O_3) could be monitored regularly in industrial zones and disclosed to the public every day. A national apex organization on AQM could require action plans setting minimum air quality standards and proposing interventions for major cities.

In several countries, command and control instruments are used in conjunction with voluntary agreements. In these countries, regulators negotiate with polluters for clean-production agreements that target either specific sectors (such as transportation or agriculture) or specific regions. The agreement involves a quid pro quo. Polluters pledge to improve environmental performance over a specified period and, in exchange, the regulator provides a grace period to allow the polluter to achieve compliance. The purpose of such agreements is to mitigate chronic noncompliance in certain sectors and regions by "building consensus" among polluters on the need for compliance and by providing guidance on how to comply. An alternative approach is "Performance-Oriented Regulations." In this approach, the regulation identifies specific environmental performance goals, such as reducing the amount of pollution associated with a process, and each facility is left to determine the best method to achieve this goal (Blackman and others 2005). One concern with this approach is that it tends to be much more difficult to enforce because it requires an intimate understanding of the alternative process (Stuart 2006).

The establishment of systematic regulatory enforcement programs for stationary sources might focus on cement, fertilizer, sugar mills, power plants, brick kilns, and textile subsectors. The main source of emissions in industry comes from the poor maintenance of boilers and generators. Brick kilns are a major source of air pollution in some areas of Pakistan. Innovative programs in other developing countries, such as Bangladesh and Mexico, can serve as models for these activities.

The courts have a significant role to play in AQM, as some international experience has showed in recent years. In India, two nongovernment organizations (NGOs), the Indian Council for Enviro-Legal Action and the Centre for Science and Environment, brought public interest lawsuits to defend local air quality. The existence of a solid constituency and free operation of the courts enabled the Supreme Court to play a decisive role in AQM. In Pakistan, the Supreme Court has taken a role in environmental policy making, particularly as it has considered several environmental degradation and protection cases. However, further efforts are needed to bring the Court's involvement with AQM to fruition.

Public Disclosure

Building environmental constituencies through dissemination of information and targeted training can be one of the essential AQM building blocks in Pakistan.

Mechanisms to disseminate information in a manner that is easily understood can empower communities to function as informal regulators. Such mechanisms also promote accountability on the part of those being regulated. Illustrative examples are the pioneering public disclosure schemes in Colombia and Indonesia, which encouraged firms to clean up their air and water pollution (Ahmed and Sánchez-Triana 2008; Blair 2008; Sánchez-Triana and Ortolano 2005; World Bank 2005).

In the early 1980s in Colombia and in the mid-1990s in Indonesia, the governments introduced schemes requiring industries to report their pollutant emissions and rate themselves on compliance with national standards. These ratings were then released publicly. Those receiving failing grades were shamed. The simplicity of the schemes allowed for easy public dissemination. The public immediately grasped the importance of factory ratings and pressured offending factories to reform. In Indonesia, factories meeting the 'green standard' were publicly praised by government officials, NGOs, and the press (Blair 2008).

Public disclosure of environmental licenses and measures to fight financial opacity of environmental policies can contribute to stimulate the demand for good environmental governance in Pakistan. Additional opportunities to strengthen environmental accountability in Pakistan would include a public information program to support clean air through the public provision of air quality information. For instance, the publication of an air quality index in major cities would build support for initiatives to improve air quality and enable the issuance of health alarms when necessary. Environmental accountability to stakeholders is essential for sound environmental management in Pakistan and will only be achieved by ensuring that stakeholders are informed and empowered. Public support for environmental compliance can be reinforced both by involving concerned civil society stakeholders in environmental decision making and oversight, and by supporting public interest advocacy through legal associations and the establishment of environmental law clinics at universities (World Bank 2006).

Cleaner Production

Cleaner production can help boost Pakistan's industrial competitiveness.[4] Pakistani companies fare poorly in international competitiveness. Companies lag in conducting formal research and developing high-technology products and advanced production processes. In general, companies do not adopt state-of-the-art technologies or spend much on research and development activities, and business collaboration with local universities is minimal at best. Local suppliers have limited technological capabilities, and hence are not able to assist in developing new products and processes (World Economic Forum 2009, 249).

Cleaner production programs aim at increasing competitiveness and the efficiency of firms because they help firms save energy, conserve water, control pollution, ensure safety of machines and equipment, improve health and safety of workers, and improve environmental conditions and the image of the firm at the local and international levels. According to Luken (2009), reducing emissions from industrial sources in Pakistan is financially viable and affordable.[5] Some cleaner technology interventions involve tuning and improving the maintenance of boilers and generators. Others involve technologies to remove PM from air emissions stacks.[6] Many large firms in Pakistan in the textile and leather sectors that are already undertaking energy efficiency measures demonstrate this (Khan 2010; Laiq 2011).[7]

Control of Diffuse Sources

Several measures and proper institutional coordination are still needed to tackle diffuse sources of air pollution in Pakistan. Solid waste collection by government-owned and government-operated services in Pakistan's cities currently averages only 50% of waste quantities generated; however, for cities to be relatively clean, at least 75% of these quantities should be collected. None of the cities in Pakistan has a proper solid-waste management system from collection of solid waste through to its proper disposal (Ghauri 2010). The recommended measures in the PCAP to address dispersed area sources are (a) block tree plantation; (b) afforestation in deserts; (c) sand dune stabilization; (d) paving of shoulders along roads; and (e) the proper disposal of solid waste. Implementing these measures will require a coordinated approach between the Ministry of Climate Change, provincial EPAs, and other departments and local governments (Ghauri 2010).

The burning sugarcane fields (to facilitate harvesting) are an important source of air pollution. Farmers in Pakistan burn cane fields to increase ease of harvesting by reducing extraneous leafy material (Khan 2010). Green cut agricultural practices can contribute to reducing nonpoint source emissions from burning out the fields for planting. In addition, dry weather conditions lead to heavy accumulation of pollutants in the atmosphere.

Conclusions and Recommendations

Stationary sources, including point and nonpoint sources, are significant contributors to air pollution in Pakistan's urban areas. Currently, data on emissions from these sources are not systematically collected. Also, the institutional framework for AQM has yet to incorporate many of the instruments that have proved effective in controlling stationary sources in other countries confronting severe air quality problems.

Developing an inventory of stationary air pollution sources is an indispensable pillar of AQM. Official data indicate that only around 1% of the country's industrial establishments report their emissions under the SMART program. As a result, there is little to no information that can help the Pak-EPA or the provincial EPAs to map the location of different sources of air pollution, assess the types and quantities of pollutants that are being discharged into the environment, identify non-compliant units, and develop the necessary corrective actions. The experiences of other countries, such as Chile and Mexico, which have developed Pollutant Release and Transfer Registers, provide important lessons for Pakistan. Under such registers, firms with an environmental license are legally mandated to report periodically their emissions of a set of health-damaging substances. In return, government agencies are legally required to protect confidential data submitted by private firms. In addition, the authority that compiles the information is required to disseminate periodic reports that can inform air quality constituents about trends in pollution emissions and associated health risks (OECD 2000).

SMART and other voluntary programs should be phased out and replaced by mandatory instruments. Even though voluntary programs may have been effective under very specific circumstances, evidence from around the world, including both developed and developing countries, indicates that voluntary programs are more ineffective for controlling pollution than mandatory instruments (Esty and Porter 2001; Morgenstern and Pizer 2007). In Pakistan, the extremely low number of units that participate in SMART evidences the tool's limitations and its ineffectiveness to generate the information and actions necessary to improve air quality. Therefore, SMART and other voluntary programs should be phased out and replaced with legal instruments that obligate polluters to comply with clear and enforceable environmental standards.

Command and control instruments can be effective to control air pollution. For air pollution control, the most common command and control measures include ambient standards, emission standards, and technology-based and performance-based standards. Pakistan could start setting and monitoring standards for PM, particularly $PM_{2.5}$, as one of the pollutants to be regularly monitored.

Market-based instruments, such as taxes or pricing policies, could be a more effective tool for moving existing markets and policies towards improved environmental outcomes. However, there is a lack of understanding between the GoP and industries regarding the application of the PPP. A weak environmental framework, especially the failure to stipulate whether industries are responsible for bearing the cost of pollution reduction, hinders the successful implementation of cleaner production systems in heavily polluting industries. In Pakistan, pollution charges could be designed to promote a shift from using highly polluting fuels such as biomass and fuel oil to cleaner fuels such as natural gas. Pollution charges could also target greenhouse gases (GHG) with the aim of mitigating climate change. Removing fuel price distortions is intended to bring domestic prices in line with international prices and eliminate the difference in tax rates among fuels. The removal of subsidies and price distortions will result in changes in the fuel mix and energy efficient outcomes.

The PPP, in which firms bear the costs of pollution reduction efforts, would enable the implementation of PCAP. The PPP is an essential pillar in reforming the AQM framework. The PPP is a global best practice and was included in the nonbinding Rio Declaration on Environment and Development of 1992 that Pakistan signed. However, in practice, no agreement or legal disposition requires polluters to pay for remedying the damages they cause to the environment (Luken 2009).

Cleaner production has the potential to support improved environmental outcomes, including reduced air pollution, and strengthen firm competitiveness. It should thus be promoted in the short term. Other measures that could be supported in the medium term to control emissions from industrial sources include requirements to minimize leaks from the transport and storage of fuels, chemicals, and materials that emit volatile organic compounds (VOCs). Emissions from generators and industrial processes, particularly cement kilns and industrial furnaces, should also be targeted in the medium term.

The courts have a significant role to play in AQM, as some international experience has showed in recent years. In countries such as India, the existence of a solid constituency and free operation of the courts enabled the Supreme Court to play a decisive role in AQM. In Pakistan, there are encouraging precedents, because the Supreme Court has taken a role in environmental policy making, particularly because it has considered several environmental degradation and protection cases. However, further efforts, as well as promoting constituency formation and the free operation of the courts, are necessary to bring the Court's involvement with AQM to fruition.

In addition, public disclosure policies can be very effective in reducing air pollution. Building environmental constituencies through dissemination of information is one of the essential AQM building blocks. The experiences of Colombia in the 1980s and Indonesia in the 1990s demonstrate how "name and shame" schemes can constitute powerful incentives for firms to reduce their pollution emissions. Given that access to niche and export markets increasingly requires that firms adopt sound environmental management practices, public disclosure policies are likely to remain relevant and effective.

Nonpoint sources, including solid waste disposal and agricultural residues, should also be included in efforts to improve air quality in Pakistan. Nonpoint sources are comparatively more difficult to monitor and tackle than point sources; however, they should not been neglected in AQM strategies, as available evidence indicates that they substantially contribute to poor air quality. The collection and disposal of solid wastes should be expanded significantly above current levels, which in many cities cover only 50% of wastes generated. The GoP should work with urban authorities to improve waste management, with the aim of significantly reducing waste burning and inadequate disposal. Short-term efforts should focus on large urban areas, with efforts increasingly targeting medium and small population centers.

In the case of agricultural residues, countries such as Brazil and Colombia offer valuable lessons that the GoP might consider. In these countries, the associations of sugarcane farmers have agreed to use the 'green cut' method, which relies on manual or mechanical removal of sugarcane, instead of burning. This method is particularly used for harvesting crops within 1,000 meters of urban areas, 30 meters from villages, 80 meters from highways, and 30 meters around high-tension transmission lines (Morgenstern and Pizer 2007).

Other measures that the GoP should consider to control air pollution from nonpoint sources in the medium to long term include paving of roads and parking areas to control PM_{10}, increasing vegetation cover to reduce dust emissions, and implementing a program to eradicate open defecation. Regulations should also be adopted to address air pollutions emissions from construction and demolition sites, such as requiring spraying of water and other substances or covering materials such as sand or cement to avoid their dispersion due to wind and other factors, as well as controlling visible dust emissions (at an opacity level of 20% or higher).

Notes

1. This chapter is based on consultant reports prepared by Burki and others (2011), Ghauri (2010), Khan (2010), and Luken (2009) for the World Bank.
2. Fly ash is produced when pith, along with bagasse, is burned in the combustion chamber. Pith (being lighter) makes its way into the atmosphere through the boiler stack. Fly ash contains carbon content in the range of 28–73%. About 3 kg of ash and fly ash is produced per tonne of cane. Fly-ash catchers and cyclone separators remove fly ash from the flue gases. Lighter particles of fly ash find their way through the boiler's stack into the atmosphere because of improper design and due to low removal efficiencies of ash removal system.
3. These variables are (a) strictness of environmental pollution standards, (b) stringency of environmental enforcement, (c) maturity of environmental regulatory structure, (d) quality of environmental information available, (e) quality of environmental institutions, and (f) the extent of subsidization of natural resource consumption, such as subsidization of fuels.
4. According to UNEP, cleaner production is "the continuous application of an integrated environmental strategy to processes, products and services to increase efficiency and reduce risks to humans and the environment" (Luken 2009).
5. At selected paper mills, implementation of energy efficient measures resulted in a 5% reduction of electricity and fuel consumption (Khan 2010). In paper manufacturing factories, cleaner production also helped eliminate chlorine emission from the process. Improvement in the bleaching process resulted in a 10% reduction in chemical consumption (Khan 2010).
6. Selected cement factories in Pakistan have installed electrostatic precipitators and bag filters to control the PM emissions. Electrostatic precipitators allow the recovery of fine cement (Khan 2010).
7. Energy efficiency in textile factories in Pakistan includes the following processes:

   - implementation of countercurrent washing throughout the production process;
   - improvement of compressed air system;
   - installation and maintenance of steam traps;
   - installation of temperature and pressure gauges, gas and steam flow meters, heat exchangers and economizers, electrical inverters on variable-speed equipment, and energy-efficient lighting and motors;
   - leakage control of steam and compressed air;
   - power factor improvements;
   - reuse of steam condensate; and
   - thermal insulation of bare hot pipelines and equipment.

References

Ahmed, K., and E. Sánchez-Triana, ed. 2008. *Strategic Environmental Assessment for Policies—An Instrument for Good Governance.* Washington, DC: World Bank. http://elibrary.worldbank.org/docserver/download/9780821367629.pdf?expires=1312689673&id=id&accname=ic_stanford&checksum=CC659EB596EA6579296944C3C7CDDA6E.

Alahdad, Z. 2011. "Turning Energy Around." In *Pakistan: Beyond the Crisis State,* edited by M. Lodhi, 231–50. New York, NY: Columbia University Press—Columbia/Hurst.

Blackman, A., R. Morgenstern, L. Montealegre-Murcia, and J. C. Garcia de Brigard. 2005. *Review of the Efficiency and Effectiveness of Colombia's Environmental Policies. Final Report to the World Bank.* Washington, DC: Resources for the Future.

Blair, H. 2008. "Building and Reinforcing Social Accountability for Improved Environmental Governance." In *Strategic Environmental Assessment for Policies: An Instrument for Good Governance,* edited by K. Ahmed and E. Sánchez-Triana, 127–57. Washington, DC: World Bank. https://openknowledge.worldbank.org/handle/10986/6461 License: Creative Commons Attribution CC BY 3.0.

Burki, A. A., K. A. Munir, M. A. Khan, M. U. Khan, A. Faheem, A. Khalid, and S. T. Hussain. 2011. *Industrial Policy, Its Spatial Aspects and Cluster Development in Pakistan.* Consultant report commissioned by the World Bank, Washington, DC.

CIA (Central Intelligence Agency). 2008. *World Factbook 2008.* https://www.cia.gov/library/publications/the-world-factbook/geos/mx.html.

Esty, D., and M. E. Porter. 2001. "Ranking National Environmental Regulation and Performance: A Leading Indicator of Future Competitiveness?" *Global Competitiveness Report 2001–2002*; New York: Oxford University Press. http://www.stadt-zuerich.ch/content/dam/stzh/prd/Deutsch/Stadtentwicklung/Publikationen_und_Broschueren/Wirtschaftsfoerderung/Standort_Zuerich/GCR_20012002_Environment.pdf.

FBS (Federal Bureau of Statistics). 2004. "Compendium of Environmental Statistics." Statistics Division. http://www.statpak.gov.pk/fbs/content/compendium-environment-statistics-pakistan-2004.

———. 2005–06. "Census of Manufacturing Industries (CMI)." http://www.statpak.gov.pk/fbs/content/census-manufacturing-industries-cmi-2005–06.

Ghauri, B. 2010. *Institutional Analysis of Air Quality Management in Urban Pakistan.* Consulting report commissioned by the World Bank, Washington, DC.

GoP (Government of Pakistan). 2011. *Framework for Economic Growth.* Islamabad: Planning Commission, GoP.

Gurjar, B. R., L. T. Molina, and C. S. P. Ojha. 2010. *Air Pollution: Health and Environmental Impacts.* Boca Raton, FL: CRC Press.

Hashmi, D. R., and M. I. Khani. 2003. "Measurement of Traditional Air Pollutants in Industrial Areas of Karachi, Pakistan." *Journal of the Chemical Society of Pakistan* 25: 103–09. http://jcsp.org.pk/index.php/jcsp/article/viewFile/1413/966.

Khan, A. U. 2010. *Industrial Environmental Management in Pakistan.* Consultant report prepared for the World Bank, Washington, DC.

Kojima, M. 2009. *Government Response to Oil Price Volatility.* Extractive Industries for Development Series 10. Washington, DC: World Bank. http://siteresources.worldbank.org/INTOGMC/Resources/10-govt_response-hyperlinked.pdf.

Laiq, A. 2011. "Evaluation of Cleaner Production Initiatives in Pakistan." Study commissioned by the World Bank, Washington, DC.

Li, L., and L. Bin. 2010. *Environmental Cost Analysis of the Relocation of Pollution-Intensive Industries Case Study: Transfer of Ceramics Industry from Foshan to Qingyang, Guangdong Province*. Research Report 2010-RR2. Singapore: Economy and Environment Program for Southeast Asia.

Luken, R. 2009. *Cleaner Production in Pakistan*. Consultant report prepared for the World Bank, Washington, DC.

Ministry of Finance. 2010. *Pakistan Economic Survey 2009–10*. Islamabad: Ministry of Finance, Government of Pakistan. http://www.finance.gov.pk/survey_0910.html.

Ministry of Petroleum & Natural Resources. 2009. *Pakistan Energy Yearbook 2009*. Islamabad: Hydrocarbon Development Institute of Pakistan.

———. 2013. *Pakistan Energy Yearbook 2012*. Islamabad: Hydrocarbon Development Institute of Pakistan.

Ministry of Textile Industry. 2009. "Textiles Policy 2009–2014." http://www.textile.gov.pk.

Morgenstern, R., and W. Pizer, eds. 2007. *Reality Check: The Nature and Performance of Voluntary Environmental Programs in the United States, Europe, and Japan*. Washington, DC: Resources for the Future Press. http://www.ctci.org.tw/public /Attachment/811211110871.pdf.

OECD (Organisation for Economic Co-operation and Development). 2000. *PRTR Implementation: Member Country Progress*. http://www.oecd.org/officialdocuments /displaydocumentpdf?cote=env/epoc(2000)8/final&doclanguage=en. Accessed on April 12, 2011.

Oil and Gas Regulatory Agency. 2011. *Notification S.R.O. (I)/2010*. Islamabad: Oil and Gas Regulatory Authority. http://www.ogra.org.pk/images/data/downloads /1278047663.pdf.

Pakistan Planning Commission. 2011. *Pakistan: Framework for Economic Growth*. Government of Pakistan. http://www.pc.gov.pk/hot%20links/growth_document _english_version.pdf.

Pak-EPA (Pakistan Environmental Protection Agency). 2010. *National Environmental Quality Standards for Ambient Air*. Government of Pakistan. http://www.environment .gov.pk/act-rules/NEQS%20for%20Ambient%20Air.pdf.

Rajput, M. U., S. Ahmad, M. Ahmad, and W. Ahmad. 2005. "Determination of Elemental Composition of Atmospheric Aerosol in the Urban Area of Islamabad, Pakistan." *Journal of Radioanalytical and Nuclear Chemistry* 266: 545–50. http://www .springerlink.com/content/p2ht755w6847072h/fulltext.pdf.

Sánchez-Triana, E. 1993. *Mecanismos financieros e incentivos económicos para la protección ambiental*. Bogotá: Friedrich Ebert Stiftung.

Sánchez-Triana, E., K. Ahmed, and Y. Awe. 2007. *Environmental Priorities and Poverty Reduction: A Country Environmental Analysis for Colombia*. Directions in Development. Environment and Sustainable Development. Washington, DC: World Bank. http:// www-wds.worldbank.org/external/default/WDSContentServer/WDSP/IB/2007/08 /13/000020953_20070813145440/Rendered/PDF/405210Env0prio101OFFICIAL0 USE0ONLY1.pdf.

Sánchez-Triana, E., and L. Ortolano. 2005. "Influence of Organizational Learning on Water Pollution Control in Colombia's Cauca Valley." *International Journal of Water Resources Development* 21 (3): 493–508.

Smith, D. J. T., R. M. Harrison, L. Luhana, C. A. Pio, L. M. Castro, M. N. Tariq, S. Hayat, and T. Quraishi. 1996. "Concentrations of Particulate Airborne Polycyclic

Aromatic Hydrocarbons and Metals Collected in Lahore, Pakistan." *Atmospheric Environment* 30: 4031–40. http://www.sciencedirect.com/science?_ob=MImg& _imagekey=B6VH3-3VWK36N-D-1&_cdi=6055&_user=145269&_ pii=1352231 096001070&_origin=&_coverDate=12%2F31%2F1996&_sk=999699976&view =c&wchp=dGLbVzW-zSkzS&md5=dd42b22ca9ad40d369679e87cab599a8&ie =/sdarticle.pdf.

Stuart, Ralph. 2006. "Command and Control Regulation." Encyclopedia of Earth. http:// www.eoearth.org/view/article/51cbed4b7896bb431f691240/.

TDAP (Trade Development Authority of Pakistan). 2008. "Export of Top Products from Pakistan." http://www.tdap.gov.pk/tdap-statistics.php.

Wasim, M., A. Rahman, S. Waheed, M. Daud, and S. Ahmad. 2003. "INAA for the Characterization of Airborne Particulate Matter from the Industrial Area of Islamabad City." *Journal of Radioanalytical Nuclear Chemistry* 25: 397–402. http:// www.springerlink.com/content/p4nr4771263013h1/fulltext.pdf.

WHO/UNEP (World Health Organization/United Nations Environment Programme). 1992. *Urban Air Pollution in Megacities of the World.* World Health Organization, United Nations Environment Programme. Oxford, U.K.: Blackwell. http://alpha .chem.umb.edu/chemistry/ch471/documents/mageetal.pdf.

World Bank. 2005. *Integrating Environmental Considerations in Policy Formulation: Lessons from Policy-Based SEA Experience.* Report 32783. Washington, DC: World Bank.

———. 2006. "Pakistan Strategic Country Environmental Assessment." World Bank, Washington, DC.

———. 2007. *Republic of Peru—Environmental Sustainability: A Key to Poverty Reduction in Peru.* Washington, DC: World Bank.

World Economic Forum. 2009. *Global Competitiveness Report 2009–2010.* Geneva, Switzerland: World Economic Forum.

CHAPTER 6

Potential Co-Benefits of Air Quality Management for Climate Change

Introduction

Pakistan is one of the countries most vulnerable to climate change, as evidenced by the 2010 floods. Climate change impacts include increased intensity of extreme weather events; potential increased intensity and frequency of drought; effects on agriculture, water resources, and ecosystems (wetlands); likely initial flooding and future drying of water resources due to glacial melt; impact on water consumption; probable damages from sea-level rise; and possible outbreak of heat-related water shortages, migration, and conflict. Impacts related to climate change on agricultural production are expected to be particularly severe for Pakistan, where it is estimated that average revenues would fall by 17% by the year 2080 (World Bank 2010).

Emissions of greenhouse gases (GHGs) from Pakistan are significantly lower than those from developed countries. However, Pakistan produces more GHG emissions per gross domestic product (GDP) than any of its neighbor countries in the South Asia region, including the major contributor, India (World Bank 2010). Pakistan can take advantage of opportunities to reduce its carbon intensity that will also help improve local air quality.

The main objective of this chapter is to identify synergies among interventions to mitigate climate change and those aiming to control air pollution. The section on "Climate Change and Air Pollution Interactions" identifies the interactions between climate change and urban air pollution, and provides an overview of GHG emissions in Pakistan and its sources. The section on "Air Quality Management and Climate Change Co-Benefits in Pakistan" identifies co-benefits from interventions for air pollution control that mitigate climate change impacts. The "Policy Options" section discusses policy options for taking advantage of potential synergies between GHG mitigation and local interventions for air quality improvement. The "Conclusions and Recommendations" section summarizes conclusions and recommendations from the analysis presented in this chapter.

Climate Change and Air Pollution Interactions

Pakistan has the highest GHG emissions intensity in South Asia. Pakistan currently emits more GHG emissions per unit of GDP than any other country in the region, including India. Perhaps more worrying is the fact that Pakistan's overall trend indicates a steady increase in emissions per unit of economic output over the last decades. In contrast, India, which remains the largest emitter in the region in absolute terms, seems to have initiated a "decoupling" of emissions and economic output. The remaining countries in the region have a GHG-emission intensity that is not even half of Pakistan's (figure 6.1).

Fossil fuel combustion accounts for more than 90% of total carbon dioxide (CO_2) emissions and for 40% of the overall GHG emissions in Pakistan. The second main contributor to overall GHG emissions is the agricultural sector, which is responsible for most of the methane and nitrous oxide emissions. Within the emissions produced by fuel combustion, the transport sector accounts for 23% of all CO_2 emissions, of which the road sector takes the largest part (97%) due to its continued dependence upon CO_2 intensive fuels. Only countries with a very small industrial and energy installed capacity show a higher contribution than the road sector to GHG emissions. Based on the use of commercial energy, in Pakistan, industry seems to account for a relatively small share of total GHG emissions (figure 6.2).

Most of the transport GHG emissions originate from road transportation. Road transportation accounts for more than 90% of transport GHG emissions. Air transportation is the second most significant contributor of CO_2, with a 7% share, compared to railways and pipeline transport, which emit only 3% and 2% of transport emissions, respectively. There were also minor GHG emissions from road infrastructure development, in particular from asphalt road paving. Halocarbons, which are among other contributors to climate change, are emitted

Figure 6.1 GHG Emissions in South Asia

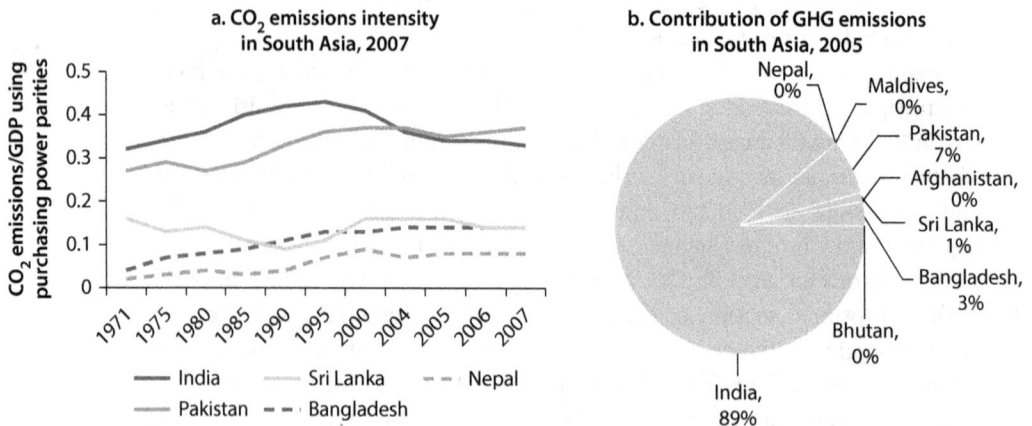

Sources: IEA 2009; World Bank 2010.

Figure 6.2 Sources of GHG Emissions in Pakistan, 2005
Percent

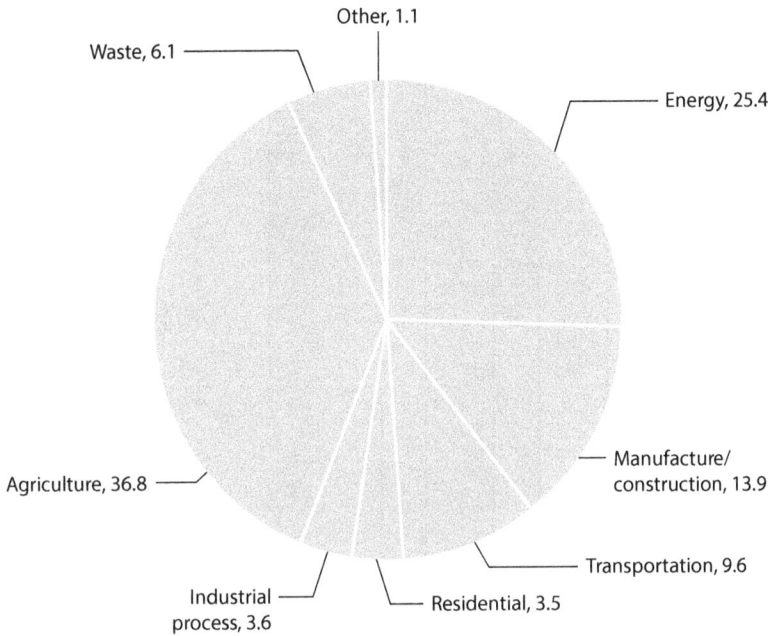

Source: World Bank calculations.

in air conditioning, which will be required in the future to relieve travelers from heat stress. During the period studied, halocarbon emissions from air conditioning and other sources amounted to 3% of GHG emissions (Ministry of Environment 2003, 33). In 2005–06, GHG emissions from road transportation amounted to 26 times the GHG emissions from railways.

The high GHG emissions from the transport sector are attributed to its continued dependence upon fuels that are CO_2 intensive. About 67% of total transport emissions are attributable to gas- and diesel-oil-operated road vehicles, according to data from 1993 to 1994. Diesel is the main fuel used for freight and passenger transportation. During 2001–02, diesel consumption amounted to 6.96 million tonnes, accounting for 84% of transport energy, and emitting about 16 million tonnes of CO_2. In 2003–04, diesel consumption went down to 6.31 million tonnes, representing 80% of total transport energy and resulting in emissions of 15 million tonnes of CO_2. Gasoline is the second largest energy source, with a share of 14% and 16% during 2001–02 and 2003–04, respectively. In 2001–12, gasoline consumption was 1.09 million tonnes and emitted 2.5 million tonnes of CO_2, and in 2003–04, its consumption and emissions rose slightly to 1.19 million tonnes and 2.7 million tonnes of CO_2. Cleaner natural gas accounts for the rest. According to Pakistan Railways, its diesel consumption was 2% of the total transport petroleum consumption in 2006–07 (figure 6.3) (Abdula 2010).

Improvements in fuel consumption and fuel mix explain the slower growth of transport CO_2 emissions from 2000 to 2010. The transport sector remained the

Figure 6.3 Sources of CO$_2$ from Fossil Fuels Combustion in Selected Asian Countries, 2007

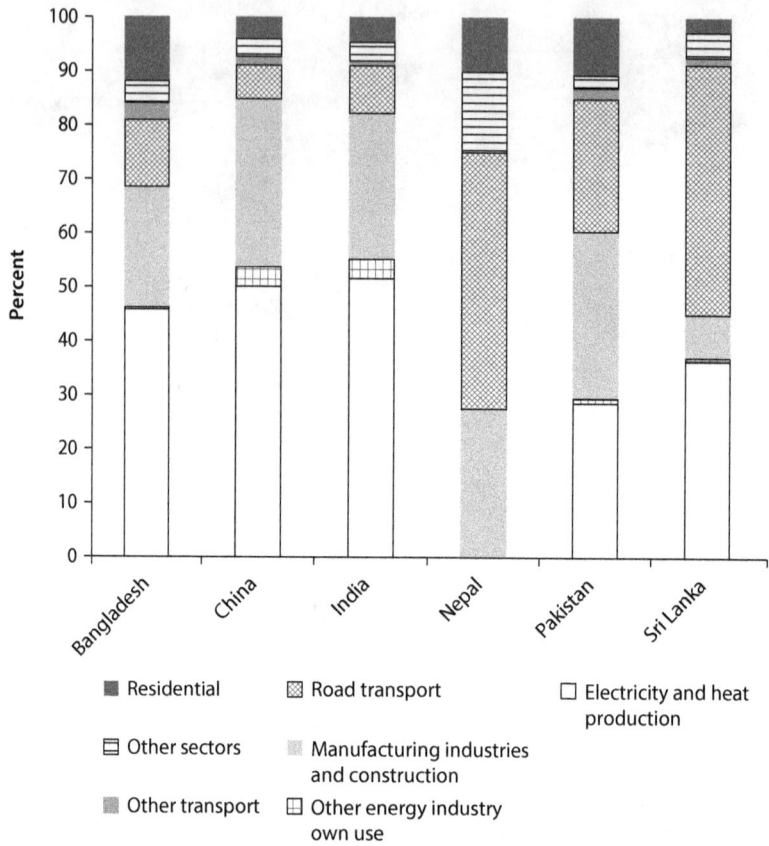

Legend:
- Residential
- Road transport
- Electricity and heat production
- Other sectors
- Manufacturing industries and construction
- Other transport
- Other energy industry own use

Source: IEA 2010b.
Note: Other energy industry own use contains emissions from fuel combusted in oil refineries, for the manufacture of solid fuels, coal mining, oil and gas extraction, and other energy-producing industries.

largest consumer of petroleum products during this period. It consumed about half of the total petroleum products during 1997–98 to 2008–09. During the period, the sector's consumption only rose by 11% and even declined by 0.2% during 2004–05 to 2008–09. The use of gas, especially in the form of compressed natural gas (CNG) in the road sector, has experienced a constant increase. Total CNG consumption increased by 84% from 1997–98 to 2006–07, although there has been a more modest growth pattern (4%) in recent years (2006–07 to 2008–09) (Hydrocarbon Development Institute of Pakistan 2010). An aggressive investment program to support conversion of motor vehicles to CNG use has driven the transition to a cleaner fuel. By 2007, more than 1.7 million vehicles operated on CNG (Ministry of Finance 2010, 246).

Air Quality Management and Climate Change Co-Benefits in Pakistan

Interventions to mitigate climate change provide co-benefits for local air quality (table 6.1). Interventions such as energy efficiency improvements, co-generation of heat and power, fuel switches from coal and oil to natural gas, and carbon capture and storage provide positive impacts on local air quality (Pleijel 2009). Efficiency improvements in the use of fossil fuels, better fuel quality, and changes in the fuel mix lead to lower emission rates of air pollutants such as PM_{10}, SO_2, and NO_x. In the European Union, implementing GHG emission-control measures available in the market would lead to 48% lower ambient concentrations of $PM_{2.5}$ in comparison to the baseline scenario (Amann and others 2008). A similar methodology applied in India suggests that the implementation of cost-efficient climate change mitigation measures would lead to a 25% reduction in health impacts associated with air pollution (Purohit and others 2009).

Black carbon particles and other air pollutants correlate with local and regional warming, as well as with health impacts. Deposition of black carbon particles alters surface light reflection of snow and ice in the Hindu-Kush-Himalaya-Tibetan glacier and contributes to atmospheric warming, which speeds the rate of glacial melt (Moore 2009). Soot and other "black" particles contribute to the formation and duration of atmospheric brown clouds, and seem to be linked to the decrease of the summer monsoon rainfall in South Asia, and the north-to-south shift of the summer monsoon in eastern China (Pleijel 2009, 15). Reducing emissions of black carbon particles provides benefits from reduction in glacier melt and pressure on local natural resources, to improved vehicle of fuel

Table 6.1 Climate Change Mitigation Interventions with Local Air Pollution Co-Benefits

Electricity generation and combustion processes
- low-carbon fuels
- solar, hydro, geothermal

Transport
- low-carbon fuels
- modal switch

Reducing black carbon emissions
- clean diesel
- natural gas

Reducing methane emissions
- flaring and leakage
- landfill
- agriculture
- water-pollution control

Reducing nitrous oxide emissions
- tune-up of industrial boilers
- use of alternate fuels

Source: Adapted from Akbar and Hamilton 2010.

efficiency (Akbar and Hamilton 2010). Even though black carbon is not part of the Kyoto Protocol (the existing international agreement aiming to control GHG emissions), it has a high warming potential; consequently, efforts should be made to mitigate its emissions.

Cleaner technologies could reduce air pollution and GHG emissions. Climate finance can make a difference in the financial feasibility of cleaner production projects. For example, while an independent power producer in Pakistan may not value the reductions in CO_2 emissions as a project benefit, the carbon-offset markets will. Selling a stream of carbon credits would therefore become part of the project benefits. If the sum of the value of the CO_2 reductions and the benefits of the project exceed that of the standard project, the emissions-reducing project becomes the preferred option for the client. If the internal rate of return of a cleaner production project is still not high enough when co-benefits and carbon credits are taken into account, then concessional financing provided by climate finance instruments may be obtained before the client is willing to invest in the project.

Improving the quality of road conditions could improve travel speeds and reduce fuel consumption and its associated emissions. The quality of road infrastructure in Pakistan has been low. Only about 60% of roads are paved nationwide, about half of its highway network is in need of major rehabilitation or reconstruction, and the remaining half will no longer be operational in the near future unless adequate and timely maintenance and rehabilitation are undertaken. In Pakistan, with poor road conditions, average travel speed is only 20–25 kph (World Bank 2006). This incurs a cost to the economy of 60–90 billion PRs per year in extra fuel costs (Abdula 2010, 5). With available international funding, the incremental costs of increasing climate resilience might become eligible for adaptation finance. Upgrading roads to become climate resilient also offers the potential of reducing traffic related emissions—with local and global benefits—for a small incremental cost.

Policy Options

In 2012, Pakistan adopted a climate change policy that incorporates a low-carbon-growth pathway. High population numbers and scarcity of land in many Pakistan cities are making urban transport systems based on private vehicles untenable, for reasons of both congestion and pollution. A broader conception of co-benefits has been proposed in the Climate Change Policy to drive transformation by including the design of cities that are greener and more livable. The intent is to substantially increase the quality of life for urban dwellers and attract more investment and businesses to cities like Karachi, Lahore, Multan, Faisalabad, Islamabad-Rawalpindi, and Peshawar—thus making those cities the engine of Pakistan's development.

Pakistan's sectors with significant climate change mitigation and air pollution control co-benefits are energy, transport, and agriculture. Energy sector policies that generate environmental co-benefits include removing fuel subsidies, fuel switching, improving energy efficiency of plants, renewable energy uptake mostly through reduction in air pollution, and improved health.

Fuel switching from coal, fuel oil, or diesel to natural gas has a potential to bring about substantial improvements in local air quality, as well as GHG mitigation. CNG emits little particulate matter, volatile organic compounds (VOCs), or SO_x when burned. Natural gas has the potential to produce less CO_2 per kilometer of travel than most other fossil fuels, and it has been estimated that natural gas vehicles can reduce GHG emissions by as much as 20–25% over gasoline vehicles (IANGV 2003, 5). The growth in CNG use in Pakistan has also provided economic benefits, in terms of employment generation and reduced dependence upon imported energy sources. The use of liquid petroleum gas (LPG) not only reduces emissions of PM and SO_x, but also CO_2 emissions. It has been estimated that this option can provide an average 30% reduction of GHG emissions per kilometer travelled (Chapman 2007, 358). The input ratio of CO_2 per input of energy unit for different fuels favors natural gas over coal, oil, furnace oil, or diesel. Data from the U.S. Energy Information Administration (1999) show that, on an input basis (pounds of CO_2 per billion Btu energy input), coal emissions (208,000) outweigh those of oil (164,000) and natural gas (117,000). The same happens with the output ratio for the whole of U.S. power generation. Coal-associated emissions (2.095–2.117 pounds CO_2/output kWh) are greater than those from petroleum (1.915–1.969 pounds CO_2/output kWh) or natural gas (1.314–1.321 pounds CO_2/output kWh).

Transport policies can be tailored in Pakistan to produce co-benefits in the form of reductions in local air pollutants and improvement in air quality, with associated health benefits and reductions in congestion, noise, and accidents. Such policies could include fostering the use of railways or modernizing the trucking sector by increasing trucks' weight. Other climate policies in the transportation sector include improving the efficiency of motorized vehicles, promotion of mass transit, policies to reduce congestion on roads and highways and in urban metropolitan centers, and promotion of non-motorized transport.

Climate policies in the agricultural sector include improved soil management practices. Due to high temperatures in the summer (40–50°C), fine dust is transported into the atmosphere with the rising hot air and forms "dust clouds" and haze over many cities of southern Punjab and upper Sindh. Dust storms are also generated from deserts (Thal, Cholistan, and Thar), particularly during the summer, and adversely affect visibility in the cities of Punjab and Sindh. Improved soil management practices increase fertility and soil quality, while at the same time enhancing adaptation to drought by improving soil water content and resource conservation.

Soil and water management enhances soil carbon sequestration (through increased organic matter residues returned to soil) and reduction of emissions from land use, land use change, and intensive agriculture practices. These actions can play an important role in the voluntary carbon markets by promoting the creation of large carbon sinks and stocks (Akbar and Hamilton 2010).

Phasing out fuel price distortions and applying the polluter pays principle can also contribute to reducing fuel consumption by promoting more efficient use of vehicles, better maintenance, and a shift to smaller vehicles.[1] In the long run, income effects derived from fuel price increases may result in a significant demand contraction, whereas cross elasticity among different energy and transport alternatives may also contribute in the same direction. These postulated effects seem to be consistent with several pieces of empirical research that have explicitly addressed the impact of fuel subsidies removals.

Nevertheless, it is important to note that the demand response would vary markedly by country, according to the extent of the pricing and the oil products concerned. Thus, as reported in the 2009 World Energy Outlook (IEA 2010a, 960), a 90% jump in the pump prices of gasoline and diesel in Indonesia in October 2005 caused consumption to drop by 13% in 2006. Removing price distortions would reduce global carbon emissions by 9%, assuming no change in world prices (Larsen and Shah 1992).

These changes have important co-benefits in reducing congestion, air pollution, and resultant health and environmental damage. Fuel price distortions carry a range of unwanted effects. Following is a summary of the most likely negative impacts (Bacon and Kojima 2006; Kojima 2009):

- Universal fuel price distortions almost always benefit high-income households more than the poor, because richer households consume more energy.
- Fuel price distortions require governments to increase their borrowing, raise additional revenue elsewhere, or reduce spending on other public goods, which in turn may result in forgone investments in more cost-effective antipoverty programs.
- If the price distortion is used to stabilize or lower final prices, it frees consumers from having to adjust their purchasing behavior to the costs of supply, instead giving them financial incentives to overconsume the subsidized commodity. This overconsumption results in deadweight loss, which is the difference between the public cost of subsidizing overconsumption minus the welfare gains of both final consumers and producers.

Conclusions and Recommendations

Climate change and air pollution are mutually reinforcing phenomena. As a result, there are important opportunities to build synergies in responses to both of these problems. Growing voluntary and compliance markets may provide financial resources for mitigating GHG emissions by undertaking actions that also result in improved urban air quality. Taking advantage of these opportunities will require aligning Pakistan's responses to both challenges and developing the institutional capacity to access such markets.

In Pakistan, fossil fuel combustion accounts for more than 90% of total CO_2 emissions and for 40% of the overall GHG emissions in the country. Most of the transport GHG emissions originate from road transport, which is also the main contributor to particulate matter emissions. Transport policies that have the potential to deliver climate change benefits and air quality benefits include fostering the use of railways, modernizing the trucking sector by increasing trucks' weight, improving the efficiency of motorized vehicles, promoting mass transit, policies to reduce congestion on roads and highways and in urban metropolitan centers, and promotion of non-motorized transport.

Pakistan's Climate Change Policy constitutes a call for a structural change in the growth pathway towards low carbon growth. The next step after the policy's approval is the operationalization of the concepts presented in the policy through a strategy to mitigate climate change. Such a strategy might include the prioritization of sectors with significant climate change mitigation and air pollution control co-benefits, particularly in the energy, transport, and agriculture sectors.

Voluntary carbon markets and the Climate Investment Funds portfolio of projects could potentially support Pakistan's efforts aimed at simultaneously reducing GHG emissions and air pollutants such as $PM_{2.5}$, lead, VOCs, O_3, SO_x, and NO_x. While firms and other economic agents in Pakistan may not value CO_2 emissions reductions, the carbon-offset markets may provide financial resources that would make the additional actions needed to achieve such mitigation an economically attractive option.

An apex organization responsible for climate change mitigation associated with air pollution control might be necessary to capitalize on potential synergies between climate change and air pollution responses. The interventions described in this chapter that could be implemented to reduce emissions of GHG and air pollutants in the energy, transportation, and agricultural sectors require intersectoral and inter-governmental coordination. As discussed in chapter 3 of this book evidence from around the world indicates that an apex body is indispensable for achieving the complex coordination that these actions require. Such an apex body is required to strengthen AQM and could play a lead role in promoting coordination in relation to Pakistan's climate change agenda, particularly regarding mitigation actions that can deliver air quality co-benefits.

Note

1. The environmental benefits achieved through the reduction of fuel subsidies could be expanded by introducing taxes on hydrocarbons in proportion to their GHG emissions. Several developed countries, including Denmark, Sweden, and Switzerland, have implemented carbon taxes to limit emissions (OECD 2008). However, the introduction of GHG taxes should be preceded by detailed analysis to assess their economic and social implications.

References

Abdula, R. 2010. "Integrating Climate Mitigation and Adaptation in the National Trade Corridor Improvement Program of Pakistan." Study commissioned by the World Bank, Washington, DC. Unpublished.

Akbar, S., and K. Hamilton. 2010. "Assessing the Environmental Co-Benefits of Climate Change Actions." World Bank Group, 2010 Environment Strategy, Analytical Background Papers, World Bank, Washington, DC.

Amann, M., I. Bertok, J. Cofala, C. Heyes, Z. Klimont, P. Rafaj, W. Schöpp, and F. Wagner. 2008. "Baseline Emission Projections for the Revision of the Gothenburg Protocol up to 2020." Background paper for the 42nd Session of the Working Group on Strategies and Review of the Convention on Long-Range Transboundary Air Pollution, Geneva, September 8–10. http://www.iiasa.ac.at/rains/reports/CIAMpercent20reportpercent 202-2008v1.pdf.

Bacon, R., and M. Kojima. 2006. *Phasing Out Subsidies: Recent Experience with Fuel in Developing Countries.* Public Policy for the Private Sector 37199. Washington, DC: World Bank. http://rru.worldbank.org/documents/publicpolicyjournal/310Bacon _Kojima.pdf.

Chapman, L. 2007. "Transport and Climate Change: A Review." *Journal of Transport Geography* 15 (5): 354–67. http://www.sciencedirect.com/science/article/pii /S0966692306001207.

Hydrocarbon Development Institute of Pakistan. 2010. *Pakistan Energy Yearbook, 2009.* Ministry of Petroleum and Natural Resources, Islamabad, Pakistan. http://www.scribd .com/doc/59661769/Pakistan-Energy-Yearbook-2009.

IANGV (International Association for Natural Gas Vehicles). 2003. "Natural Gas Vehicles and Climate Change: A Briefing Paper." http://www.consenseus.org/downloads/altfuels /Briefing_paper.pdf.

IEA (International Energy Association). 2009. *World Energy Outlook 2008.* Paris: IEA. http://www.iea.org/textbase/nppdf/free/2008/weo2008.pdf.

———. 2010a. *World Energy Outlook 2009.* Paris: IEA. http://www.iea.org/textbase /nppdf/free/2009/weo2009.pdf.

———. 2010b. *CO_2 Emissions from Fuel Combustion. 2009 Edition.* Paris: IEA. http://ccsl .iccip.net/co2highlights.pdf.

Kojima, M. 2009. *Government Response to Oil Price Volatility.* World Bank, Extractive Industries for Development Series #10. http://siteresources.worldbank.org /INTOGMC/Resources/10-govt_response-hyperlinked.pdf.

Larsen, B., and A. Shah. 1992. "World Fossil Fuel Subsidies and Global Carbon Emissions." Background paper prepared for World Development Report 1992, World Bank, Washington, DC.

Ministry of Environment. 2003. *Pakistan's Initial National Communication on Climate Change.* Islamabad: Ministry of Environment, Government of Pakistan. http://unfccc .int/resource/docs/natc/pakncl.pdf.

Ministry of Finance. 2010. *Pakistan Economic Survey 2009–10.* Islamabad, Pakistan: Ministry of Finance, Government of Pakistan. http://www.finance.gov.pk/survey _0910.html.

Moore, F. C. 2009. "Climate Change and Air Pollution: Exploring the Synergies and Potential for Mitigation in Industrializing Countries." *Sustainability* 1: 43–54. http:// www.mdpi.com/2071-1050/1/1/43/pdf.

OECD (Organisation for Economic Co-operation and Development). 2008. *Promoting Sustainable Consumption: Good Practices in OECD Countries.* Paris: OECD. http:// www.oecd.org/greengrowth/40317373.pdf.

Pleijel, H., ed. 2009. *Air Pollution & Climate Change. Two Sides of the Same Coin?* Stockholm: Swedish Environmental Protection Agency. http://swedishepa.com /upload/english/03_state_of_environment/air/air-climate/air-climate-chapter-07.pdf.

Purohit, P., M. Amann, R. Mathur, I. Gupta, S. Marwah, V. Verma, I. Bertok, J. Borken, A. Chambers, J. Cofala, C. Heyes, L. Hoglund, Z. Klimont, P. Rafaj, R. Sandler, W. Schopp, G. Toth, F. Wagner, and W. Winiwarter. 2009. *Scenarios for Cost-Effective Control of Air Pollution and Greenhouse Gases in India.* Laxenburg, Austria: International Institute for Applied Systems Analysis (IIASA). http://www.iiasa.ac.at /rains/reports/Asia/IR-GAINS-India.pdf.

U.S. Energy Information Administration. 1999. *Natural Gas 1998 Issues and Trends.* Washington, DC: U.S. Department of Energy. http://www.eia.gov/pub/oil_gas /natural_gas/analysis_publications/natural_gas_1998_issues_trends/pdf/it98.pdf.

World Bank. 2006. *Pakistan. Transport Competitiveness in Pakistan. Analytical Underpinnings for the National Trade Corridor Improvement Program.* Washington, DC: World Bank.

———. 2010. *World Bank Development Indicators.* Washington, DC: World Bank. http:// data.worldbank.org/indicator.

CHAPTER 7

Conclusions and Recommendations

Conclusions

Pakistan is the most urbanized country in South Asia, with more than 35% of the population living in urban areas, and most of them in cities of more than 1 million inhabitants. Pakistan's economy is undergoing a structural and spatial transformation leading to increasing contributions of the industrial and services sector to gross domestic product (GDP) and an urbanization process. Manufacturing industries and service companies tend to concentrate in a few geographic locations where the following resources enhance their competitiveness: the availability of specialized labor, inter-industry spillovers, higher road density, local transfer of knowledge, and access to international supplier and buyer networks. These factors largely explain the clustering of large-scale manufacturing and high associated employment levels around the metropolitan areas of Karachi and Lahore. As recognized by the Government of Pakistan (GoP), industrial growth and urbanization offer opportunities to support sustained economic growth in Pakistan. However, the potential associated with these opportunities might increase environmental degradation, particularly from air pollution. For Pakistan's city dwellers, air pollution is becoming the most significant environmental problem and is likely to worsen in the future unless targeted interventions are carried out in the short, medium, and long term.

Main sources of air pollution in Pakistan include mobile sources, such as heavy-duty vehicles and motorized 2–3 wheelers; stationary sources, such as power plants and burning of waste; and natural dust. Based on available information (which represents a partial inventory of possible emission sources), the road transport sector is responsible for 85% of particulate matter of less than 2.5 microns ($PM_{2.5}$) total emissions and 72% of particulate matter of less than 10 microns (PM_{10}) emissions. The number of vehicles in Pakistan has increased from around 2 million to 10.6 million over the last 20 years. From 1991 to 2012, the number of two-stroke engine vehicles has grown more than tenfold. In addition, emissions linked to burning fossil fuels (for running electric power plants and various industrial operations) are collectively linked to pollution in the form of particulates, as well as NO_x and SO_x, and carbon monoxide (CO). The limited data that are available suggest that ambient concentrations of health-damaging particulate matter (PM) in Pakistan are on average more than four times above levels recommended in World Health Organization guidelines.

Declining governmental attention to air quality management resulted in a significant paucity of reliable air quality data. Comparative risk assessments typically confirm that air pollution generates severe local impacts, particularly on human health. However, attention devoted to local pollution problems has declined rapidly in Pakistan to give precedence to climate change mitigation and other problems that have global impacts. From 2006 to 2009, Japanese International Cooperation Agency assisted the GoP in the design and installation of a network for monitoring air quality in five cities. Administrative and budget problems of environmental protection agencies led to inadequate operation and maintenance of the network for monitoring air quality. There was neither analysis nor disclosure of the collected data, and the monitoring of $PM_{2.5}$ concentrations was infrequent.

Existing monitoring programs have proven unable to provide information on emission of pollutants by stationary sources in Pakistan. In order to monitor air pollutants from industrial sources, the GoP established the voluntary Self-Monitoring and Reporting Tool (SMART). The level of reporting on air pollutant emissions is low, as only 99 out of 6,417 industrial facilities have registered their emissions under the SMART program. The ineffectiveness of the SMART program to disclose emissions of air pollutants in Pakistan is similar to those of voluntary programs implemented in other countries, including in the European Union and in the United States (Blackman and others 2005; Esty and Porter 2001; Morgenstern and Pizer 2007).

The Climate Change Division (CCD), the Pakistan Environmental Protection Agency (Pak-EPA) and the provincial EPAs have been successful in incorporating environmental concerns in most national-level policies and sensitizing development ministries and provincial governments to environmental issues. As part of its accomplishments in air pollution control, the GoP phased out lead from gasoline. However, the GoP has struggled with the implementation of environmental programs and policies and with the formulation of new policies. The Pakistan Environmental Protection Council (PEPC), which is the apex decision-making body on environmental issues in the country, has remained almost nonfunctional for most of its tenure. Not only has PEPC failed to meet at least twice every year as required under section 3 of the 1997 Pakistan Environmental Protection Act (PEPA) it failed to meet at all between 2004 and 2009.

The 1997 PEPA established a comprehensive framework for environmental management that includes a number of major environmental management entities. Currently, the Pak-EPA is an executive agency under the CCD. The 18th Constitutional Amendment has empowered provincial EPAs to take care of most of the environmental issues in the provinces, while the resource-constrained Pak-EPA's main responsibilities have been limited to assisting provincial governments in the formulation of rules and regulations under PEPA 1997. Provincial EPAs have been created in Punjab, Sindh, Khyber Pakhtunkhwa, Balochistan,

Gilgit Baltistan, and Azad Jammu and Kashmir (AJK). The environmental tribunals and magistrates created under PEPA have the power to hear environment-related cases and impose sanctions (for example, monetary fines and prison sentences) for noncompliance with environmental requirements. Penalties are rare, however, as most firms formally charged with noncompliance to PEPA implement required environmental measures.

The 2010 National Environmental Quality Standards (NEQS) for motor vehicle exhausts and noise include a set of emission standards for all new and in-use vehicles. These categories are subdivided into diesel (light and heavy) and petrol/gasoline-powered vehicles (passenger cars, light commercial vehicles, rickshaws, and motorcycles). For heavy diesel engines and large-goods vehicles (both locally manufactured and imported), the standard for PM is 0.15 g/Kwh, to be enforced after July 1, 2013. The smoke, carbon monoxide, and noise standards for in-use vehicles have been effective since the regulation was adopted at the end of 2010. A revision of these standards may become necessary to ensure their full alignment with international vehicle emissions standards (such as an equivalent Euro standard) and related testing procedures for new and in-use vehicles, including motorcycles, scooters, and three-wheelers.

In spite of a significant expansion of environmental regulatory powers with passage of PEPA and the issuance of Pakistan's National Environmental Policy in 2005, significant implementation gaps exist. While the NEQS and other aspects of Pakistan's environmental approach are comparable to those in other South Asian nations, they suffer because implementation and enforcement are weak. Neither the organizational structure of CCD nor that of Pak-EPA has a specific unit or department that is responsible for air quality management (AQM). To address priority problems involving air pollution, Pak-EPA has proposed establishing units specialized in air quality management at the national and provincial levels. The Pak-EPA would take over responsibilities for coordinating, designing, and implementing air quality policies. Technical cells at the national and provincial agencies would be responsible for monitoring ambient-air quality, and mobile, stationary, and diffuse emissions. Additionally, other technical cells would be responsible for regulatory enforcement and compliance.

Pakistan's ability to enjoy the economic benefits of avoiding the external costs of increased urbanization and industrialization will be seriously hampered unless the GoP takes steps to enhance efforts to control air pollution. Industrial expansion and urbanization without attention to environment will lead to increased GDP figures that are inflated by the high costs of environmental externalities.

Recommendations

Policy makers in Pakistan face an array of obstacles—including limited financial, human, and technical resources—and can only pursue a small number of strategic interventions on AQM at the same time. In the short term, in Pakistan, the primary emphasis should be on reducing levels of pollutants linked to high morbidity and mortality: $PM_{2.5}$ (and precursors like sulfur oxides [SO_x] and nitrogen oxides [NO_x]) from mobile sources. A second level of priority could be given to $PM_{2.5}$, SO_x, and emissions of lead (and other toxic metals) from stationary sources. A third level of priority could be given to ozone and its precursors, NO_x and VOCs (especially air toxics like benzene). A fourth level of priority could be given to other traditional air pollutants, such as carbon monoxide and green-house gases.

The CCD, the Pak-EPA, and the provincial EPAs could be strengthened to play a leadership role in minimizing the negative external effects of the expected increase in industrialization and urbanization by helping to move forward Pakistan's efforts in air pollution control. There is no reason why the lack of attention of Pakistan's firms to air pollution control must proceed as it has in the past.

To allocate attention to priority environmental problems such as air pollution, Pakistan can emulate most countries in the world that currently have a strong apex central environmental ministry or agency with a number of technical and action-oriented agencies designating and implementing public policies, and enforcing regulations.

A strengthened national system of environmental agencies coordinated by an apex organization at the national level could expand and enhance the reliability of the network for monitoring air quality. In order to build a robust database of air quality data, the GoP must also assign resources for developing an inventory of mobile, industrial, and stationary sources. Reliable air quality data would allow the GoP to model the transport, dispersion, and fate of air pollutants, and make decisions on the sequencing of reforms and interventions for air pollution control. Reliable data also would allow the GoP to effectively monitor and evaluate the effectiveness and economic efficiency of the PCAP.

As mentioned above, mobile sources are the largest contributor of air pollutants, particularly PM, in large urban centers in Pakistan. In the short term, the GoP might focus its efforts on controlling emissions from these sources. Options with favorable benefit-cost ratios to control air pollution from mobile sources in Pakistan include

- improving fuel quality by reducing the sulfur content in diesel;
- converting diesel minibuses and city delivery vans to compressed natural gas (CNG);
- installing diesel oxidation catalysts (DOCs) on existing large buses and trucks;
- converting existing two-stroke rickshaws to four-stroke CNG engines; and
- introducing low-sulfur fuel oil (1% sulfur) to major users located in Karachi.

Estimated health benefits per dollar spent (that is, the benefit-cost ratio) on cleaner diesel are in the range of US\$1–1.5 for light-duty diesel vehicles and US\$1.5–2.4 for large buses and trucks for both 500-ppm and 50-ppm diesel.

Benefits per rupee spent on retrofitting in-use diesel vehicles with a DOC, once 500-ppm diesel is available, are estimated at PRs 1–1.3 for large buses and trucks used within the city, but less than cost for minibuses and light-duty vans. However, the benefits per rupee spent on converting diesel minibuses and light-duty vans to CNG are estimated at PRs 1.2–1.7. The benefits of converting two-stroke rickshaws to four-stroke CNG are about twice as high as the conversion cost, and the benefits of four-stroke motorcycles (instead of two-stroke) are nearly three times higher than the additional cost of a four-stroke engine.

While the interventions discussed above can substantially reduce PM emissions from road vehicles in Karachi, there must also be consideration of interventions to control PM from other sources. These interventions include

- effectively enforcing the ban on burning solid waste in the city;
- moving existing brick kilns, metal foundries, and scrap smelters out of the city;
- considering wind directions and future urban development when deciding on acceptable locations;
- improving street cleaning to reduce resuspension of road dust; implementing measures to reduce construction dust; and
- controlling emissions from large point sources (ensuring that existing control equipment is operating properly and installing new control equipment if none is currently in place).

In the medium term, the GoP might consider adopting demand-side management measures to reduce the country's motorization trend. Key among these measures would be the development of mass transportation in Pakistan's main cities. Experiences from countries such as Brazil, Colombia, and Mexico demonstrate the benefits of relatively new public transport systems, such as Bus Rapid Transit, which can use bus-based technologies to transport increasingly larger volumes of customers at moderately high speeds even in very congested urban areas. While still substantial, the investments needed to develop and operate these systems are significantly lower than those of traditional mass transport systems, such as the underground metros. In addition, these systems have been able to demonstrate their contributions to reduce congestion and pollution, and some of them have even received international funding for their role in reducing greenhouse gas (GHG) emissions from mobile sources.

Additional policies that are worth assessing in the medium term include traffic control, restricted circulation of private cars during high pollution episodes, urban planning and land use, establishment of high occupancy vehicle lanes, measures to improve traffic flow such as 'green wave' coordination of traffic signals, and improvement of infrastructure, such as paving of roads and regular sweeping. City governments would be responsible for most of these measures, and thus capacity-building efforts would be needed targeting key cities to strengthen their capacity for urban planning and transportation management. In addition, the GoP might consider exploring options to promote the adoption of modern technology spark-ignition engines.

Natural gas emits less PM, volatile organic compounds, or sulfur oxides when burned than diesel. Pakistan has become the leading user of natural gas in the world, with 1.7 million CNG-operated vehicles in the country—about 18% of the total automotive fleet. However, mobile sources have only a 4.4% share in the total gas consumption in the country. Because gasoline is taxed much more than diesel, resulting in a much higher retail price for gasoline than diesel, CNG has substituted for gasoline rather than diesel in light-duty vehicles.

Energy prices are not just low because of fuel price distortions. They are also too low because they fail to account for the adverse health effects of fossil fuels; as a result, there is excessive use of highly polluting fossil fuels. Numerous examples demonstrate the perverse incentives linked to price distortions for energy; for example, disincentives to switch from high sulfur content diesel to CNG. Another illustration involves the disincentive to energy conservation created by heavily subsidized prices for gas to households and fertilizer plants. More generally, removal of subsidies for fuels consumed by motor vehicles and industries could help reduce demand for fuels and improve air quality. If the differences in tax rates for fuel oil and natural gas were eliminated (such as subsidies for natural gas as feedstock to fertilizer plants), domestic prices for those fuels could be brought in line with international prices. Natural gas would then become more competitive and there would be incentives for Pakistani firms to engage in international trade of natural gas, including imports from the Islamic Republic of Iran or Turkmenistan.

In terms of industrial sources, the GoP might consider modifying its environmental management approach based on the current form of the NEQS. These standards should be revised so that they are realistically attainable under conditions that currently exist in Pakistan. The NEQS, even in revised form, will not bring down levels of pollution unless firms and municipalities have incentives to reduce the mass flow rates of pollution discharges. A two-element approach can accomplish this. One element involves reviving the system of pollution charges (authorized by the PEPA) and restructuring it such that firms pay charges based on both concentration levels and volume flow rates of discharges. Such charges will give firms an incentive to reduce both the concentrations and masses of their air pollutant loads. The second element involves the distribution of revenues generated by pollution charges. Part of the revenue can be directed to supporting environmental protection agencies, thereby giving agency staffs incentives to monitor air pollutant emissions and collect pollution charges. Once high-priority pollutants are brought under control, the NEQS can be made more stringent for pollutants that are relatively less damaging to human health under current conditions. By proceeding in this way, Pakistan will be able to make progress on its highest priority pollution problems and allow firms to meet NEQS, which is a condition for obtaining internationally recognized environmental certifications, such as ISO 14001.

Pakistan's sectors with significant climate change mitigation and air pollution control co-benefits are energy, transport, and agriculture. Energy sector policies that generate environmental co-benefits include removing fuel price distortions, applying the polluter pays principle, fuel switching, improving energy efficiency of plants, and renewable energy uptake. Climate policies in the transportation sector include improving the efficiency of motorized vehicles and the transportation system; promotion of mass transit; policies to reduce congestion on roads, highways, and in urban metropolitan centers; and promotion of non-motorized transport.

In sum, addressing Pakistan's severe urban air pollution problem will require undertaking a series of coordinated interventions to strengthen air quality monitoring, build the institutional capacity of responsible agencies, bolster the legal and regulatory framework for AQM, carry out targeted policy reforms and investments, and fill existing knowledge gaps. Table 7.1 summarizes key recommendations emerging from this book.

Table 7.1 Recommended Actions to Strengthen Air Quality Management in Pakistan

Recommended action	Timeframe
Strengthening air quality monitoring	
Establish a reliable air quality-monitoring network focusing on pollutants such as $PM_{2.5}$, SO_2, NO_2, and lead and other toxic substances; building on the network developed with JICA support; and providing ongoing training and budget for maintaining and utilizing the equipment.	Short term
Develop a detailed mobile source emissions inventory.	Short term
Establish an inventory of industrial sources, focusing on key polluters and evolving to include small and medium enterprises.	Short term
Establish a centralized depository to review and analyze data collected from across the country by the network for monitoring air quality.	Medium term
Carry out modeling efforts to assess the present and future contributions of mobile, stationary, nonpoint, and natural sources of key pollutants.	Medium term
Building institutional capacity for AQM	
Establish a central apex organization with a clear mandate for AQM and with responsibilities for intersectoral and intergovernmental (national and provincial) coordination.	Short term
Strengthen provincial EPAs to carry out monitoring, enforcement, and planning activities related to AQM.	Short term
Strengthen enforcement of air quality regulations, and the capacity of the legal courts and judiciary to enforce environmental laws.	Short term
Build the capacity of city governments to improve urban planning and develop alternative modes of transportation.	Medium term
Bolstering the legal and regulatory framework for AQM	
Adopt pollution charges targeting fuels according to their pollution contributions, based on the polluter pays principle.	Short term
Adopt a more serious penalty system for noncompliance with air quality laws and regulations.	Short term
Phase out voluntary regulatory schemes and introduce enforceable standards for key stationary sources of air pollutants.	Short term
Adopt higher fuel efficiency standards and environmental taxes for less efficient vehicles.	Medium term
Adopt revised NEQS to ensure that the permissible ambient concentrations of all pollutants are consistent with the levels recommended by the WHO.	Medium term
Adopt taxes on hydrocarbons, based on their contribution to greenhouse gas emissions.	Long term

table continues next page

Cleaning Pakistan's Air • http://dx.doi.org/10.1596/978-1-4648-0235-5

Table 7.1 Recommended Actions to Strengthen Air Quality Management in Pakistan (continued)

Recommended action	Timeframe
Policy reforms and investments for improved air quality	
Reduce fuel price distortions and apply the polluter pays principle to promote efficient fuel use and switching towards gas and other cleaner fuels.	Short term
Improve fuel quality by importing cleaner diesel and furnace oil, using low-sulfur crude oil, and investing in refinery capacity for desulfurization.	Short term
Increase solid waste collection and proper disposal, particularly in large urban areas.	Short term
Implement green-cut agricultural practices to reduce emissions from nonpoint sources.	Short term
Promote cleaner production in the industrial sector with the dual aim of improving environmental conditions and strengthening firms' competitiveness.	Short term
Foster the creation of a strong air quality constituency by providing training and disseminating specific materials among policy makers, legislators, NGOs, journalists, and other stakeholders.	Short term
Initiate a vehicle inspection and maintenance program that initially targets diesel trucks and buses, focusing on Pakistan's larger cities, and concentrating on controlling PM emissions.	Medium term
Phase out two-stroke vehicles in favor of four-stroke vehicles, taking into account the experiences of other Asian countries.	Medium term
Require the use of catalytic converters for vehicles, particularly for gasoline-fueled vehicles.	Medium term
Adopt a public disclosure scheme requiring industries to report their pollutant emissions and rate themselves on compliance with national standards.	Medium term
Publish an Air Quality Index in major cities and issue health alarms when necessary.	Medium term
Support development of mass transportation and non-motorized vehicles in main cities.	Medium term
Relocate industries from areas where ambient air pollution concentrations exceed the legal limits to places with no thermal inversions, good atmospheric mixing conditions, and outside areas of high population density.	Medium term
Develop and implement transport policies that reduce local air pollution, congestion, noise, accidents, and have climate change mitigation co-benefits.	Medium term
Develop and implement agricultural policies, including soil management practices that increase soil quality and enhance carbon sequestration.	Medium term
Reduce emissions from nonpoint sources through measures such as paving roads and parking areas, increasing vegetation cover, and controlling emissions from construction and demolition sites.	Medium term
Control emissions from industrial sources, particularly from the storage and transportation of fuels and chemicals, as well as from high-emission industrial processes.	Medium term
Adopt fuel efficiency and emissions standards for vehicles in line with Euro 2 or 3 and gradually tighten them in line with Euro 5 or 6.	Medium to long term
Adopt a roadside inspection program to identify polluting vehicles.	Long term
Filling knowledge gaps for AQM	
Carry out an in-depth study to identify sources of pollution for lead and other toxic substances to serve as a basis for the development of targeted interventions.	Short term
Conduct in-depth analysis of a potential scrappage program for old vehicles.	Short term
Promote the establishment of research programs in universities with a focus on the different areas of AQM, including law, economics, meteorology, chemistry, and so forth.	Short term

Note: AQM = air quality management; JICA = Japanese International Cooperation Agency; NEQS = National Environmental Quality Standards; NGO = nongovernmental organization; $PM_{2.5}$ = particulate matter of less than 2.5 microns; SO_2 = sulfur dioxide; NO_2 = nitrogen dioxide.

References

Blackman, A., R. Morgenstern, L. M. Murcia, and J. C. Garcia de Brigard. 2005. *Review of the Efficiency and Effectiveness of Colombia's Environmental Policies. Final Report to the World Bank*. Washington, DC: Resources for the Future. http://www.rff.org/rff /documents/rff-rpt-coloepefficiency.pdf.

Esty, D., and M. E. Porter. 2001. "Ranking National Environmental Regulation and Performance: A Leading Indicator of Future Competitiveness?" In *Global Competitiveness Report 2001–2002*; New York: Oxford University Press. http://www .stadt-zuerich.ch/content/dam/stzh/prd/Deutsch/Stadtentwicklung/Publikationen _und_Broschueren/Wirtschaftsfoerderung/Standort_Zuerich/GCR_20012002 _Environment.pdf.

Morgenstern, R., and B. Pizer. 2007. *Reality Check. The Nature of Performance of Voluntary Environmental Programs in the United States, Europe, and Japan*. Washington, DC: Resources for the Future.

Environmental Benefits Statement

The World Bank Group is committed to reducing its environmental footprint. In support of this commitment, the Publishing and Knowledge Division leverages electronic publishing options and print-on-demand technology, which is located in regional hubs worldwide. Together, these initiatives enable print runs to be lowered and shipping distances decreased, resulting in reduced paper consumption, chemical use, greenhouse gas emissions, and waste.

The Publishing and Knowledge Division follows the recommended standards for paper use set by the Green Press Initiative. Whenever possible, books are printed on 50 percent to 100 percent postconsumer recycled paper, and at least 50 percent of the fiber in our book paper is either unbleached or bleached using Totally Chlorine Free (TCF), Processed Chlorine Free (PCF), or Enhanced Elemental Chlorine Free (EECF) processes.

More information about the Bank's environmental philosophy can be found at http://crinfo.worldbank.org/wbcrinfo/node/4.

green
press
INITIATIVE

www.ingramcontent.com/pod-product-compliance
Lightning Source LLC
Chambersburg PA
CBHW080421270326
41929CB00018B/3116